Lecture Notes in Mathematics

Edited by A. Dold and B. Eckmann

889

Jean Bourgain

New Classes of $\mathcal{L}^p$-Spaces

Springer-Verlag
Berlin Heidelberg New York 1981

Author

Jean Bourgain
Department of Mathematics, Vrije Universiteit
Pleinlaan 2-F7, 1050 Brussels, Belgium

AMS Subject Classifications (1980): 46B20, 46B25, 46E30

ISBN 3-540-11156-5 Springer-Verlag Berlin Heidelberg New York
ISBN 0-387-11156-5 Springer-Verlag New York Heidelberg Berlin

Printing and binding: Beltz Offsetdruck, Hemsbach/Bergstr.
2141/3140-543210

INTRODUCTION

The purpose of this text is to present new examples of $\mathcal{L}^p$-spaces for $1 \leq p \leq \infty$. As main reference about $\mathcal{L}^p$-theory, we mention J. Lindenstrauss and L. Tzafriri's book "Classical Banach spaces" (see [87]). The reader may also consult related papers for further details and results.
Let us start with the definition of a $\mathcal{L}^p$-space. For $1 \leq p \leq \infty$, we denote $l^p(n)$ the vectorspace $\mathbb{R}^n$ equipped with the l^p-norm.
A $\mathcal{L}^p$-space is a Banach space with $l^p(n)$-local structure. More precisely, X is a $\mathcal{L}^p$-space $(1 \leq \lambda < \infty)$ provided any finite dimensional subspace E of X is contained in a finite dimensional subspace F of X which is up to λ isomorphic to $l^p(\dim F)$.
As a result of isometric Banach space theory, we know that a Banach space is a $\mathcal{L}^p_\lambda$-space for all $\lambda > 1$ if and only if it is an $L^p(\mu)$-space (or its dual an $L^1(\mu)$-space if $p = \infty$).
Our interest goes here to the isomorphic theory. Although the definition of $\mathcal{L}^p$-spaces is finite dimensional, it has various implications on the global structure of the space and $\mathcal{L}^p$-spaces have several nice infinite dimensional properties. However, there are examples of $\mathcal{L}^p$-spaces which are not L^p-spaces. In fact, our work shows that there exist $\mathcal{L}^p$-spaces which are "pathological" in many aspects and that we may not hope for a simple description as in the isometric case.
The complemented subspaces of the classical Banach spaces, i.e. C(K)- and $L^p(\mu)$-spaces, are still not characterized and this problem is probably the most important in isomorphic Banach space theory. It can be shown that if a Banach space is complemented in a C(K)- (resp. $L^1(\mu)$-) space, then it is a $\mathcal{L}^\infty$ (resp. $\mathcal{L}^1$) space. Complemented subspaces of L^p-spaces $(1 < p < \infty)$ are either Hilbert-spaces or $\mathcal{L}^p$-spaces.
Conversely, for $1 < p < \infty$, any $\mathcal{L}^p$-space is isomorphic to a complemented subspace of an L^p-space. This is not the case for $p = 1$ or $p = \infty$. The only complemented subspaces of $L^1(\mu)$- (resp. C(K)-) spaces are conjectured to be again $L^1(\mu)$- (resp. C(K)-) spaces.

We will give new examples of $\mathcal{L}^\infty$-spaces (chapter III), $\mathcal{L}^p$-spaces for $1 < p < \infty$ (chapter IV) and $\mathcal{L}^1$-spaces (chapter V).

The $\mathcal{L}^{\infty}$-examples solve several old conjectures in $\mathcal{L}^{\infty}$-theory negatively. They also show that one may not hope for an ordinal classification of $\mathcal{L}^{\infty}$-spaces as for C(K)-spaces.
The examples of $\mathcal{L}^{p}$-spaces for $1 < p < \infty$ and $\mathcal{L}^{1}$-spaces are related and are constructed using trees on the integers. The main result here is to show that for $1 \leq p < \infty$ there exists a system of $\mathcal{L}^{p}$-spaces between l^p and L^p for which L^p is the only universal element. For $1 < p < \infty$, these spaces are obtained as complemented "tree-subspaces" R^p_α ($\alpha < \omega_1$) of L^p. We prove that for $p = 1$ the spaces R^1_α are proper subspaces of certain operators on L^1 not fixing a copy of L^1 and this allows us to construct for each ordinal $\alpha < \omega_1$ a $\mathcal{L}^{1}$-space containing R^1_α and not containing L^1.

We think that this work has three reasons of interest.
First, of course, it provides new $\mathcal{L}^{p}$-spaces. Secondly, because certain constructions are based on new ideas and techniques with possibly other applications. Finally, especially in chapter IV and chapter V, crucial use is made of certain probabilistic results which have an independent importance.

Part of this work was done jointly with F. Delbaen at the Free University of Brussels and with H.P. Rosenthal and G. Schechtman during a visiting stay at the Paris VI University and the Ecole Polytechnique. Several of these results were the subject of seminar talks given at latter institute.

I am indebted to L. Guéry-Straetmans who carried out the task of typing the manuscript.

Brussels, 4/12/1979

J. BOURGAIN

TABLE OF CONTENTS

NEW CLASSES OF $\mathcal{L}^p$-SPACES
$(1 \leq p \leq \infty)$

J. BOURGAIN

I. PRELIMINARIES

The purpose of this first chapter is to introduce the $\mathcal{L}^p$-spaces and discuss their basic properties. For complete proofs and more details we also refer the reader to [86] and [87].

1. SOME TOPOLOGICAL PROPERTIES OF BANACH SPACES

Our aim here is to recall certain notions of general Banach space theory and the relations between these notions. They will have their importance especially in the study of $\mathcal{L}^1$ and $\mathcal{L}^\infty$-spaces.
For simplicity, we only consider real Banach spaces.

A sequence (x_n) in a Banach space X is called a weak Cauchy sequence provided $(x^*(x_n))$ converges, whenever x^* is a member of the dual X^* of X.
Let A be a subset of X. We say that A is weakly conditionally compact (WCC) if every sequence in A has a weak Cauchy subsequence.
Since a weak Cauchy sequence is always bounded, WCC sets are bounded.

We say that a bounded sequence (x_n) in X is equivalent to the usual ℓ^1-basis, if there exists $\delta > 0$ such that

$$\|\Sigma_{k=1}^n a_k x_k\| \geq \delta \Sigma_{k=1}^n |a_k|$$

whenever $a_1,\ldots,a_n$ is a finite set of scalars.
It is clear that then the closed subspace $[x_n ; n]$ of X generated by the sequence (x_n) is isomorphic to ℓ^1.

It is well-known (and easily seen) that if (x_n) is an ℓ^1-basis in X, then (x_n) has no weak Cauchy subsequences. In fact, the following holds

PROPOSITION 1.1 :
1. A bounded sequence (x_n) in a Banach space X either has a weak Cauchy subsequence or has a subsequence which is equivalent to the usual ℓ^1-basis.
2. Consequently, any bounded non-WCC subset of X contains an ℓ^1-basis.

This result, due to H.P. Rosenthal [104], is of settheoretical nature.

An easy proof is obtained by using Ramsey type arguments.
Let us remark that prop. 1.1 also holds in complex version, as shown by L. Dor [46].

WCC sets have several nice stability properties. So for instance

PROPOSITION 1.2 : If A is a WCC subset of X, then also the absolutely convex hull $\Gamma(A)$ of A is WCC.

Related results can be found in [94], [105] and [33].

If X and Y are Banach spaces, then we say that X is hereditarily Y provided every infinite dimensional, closed subspace of X contains an isomorphic copy of Y.
If X, Y and Z are Banach spaces and $T : X \to Y$ an operator, then we say that T fixes a copy of Z provided T is an into isomorphism when restricted to some subspace of X which is isomorphic to Z.

The next proposition is a more or less direct consequence of Eberlein's theorem on weak compactness and the preceding

PROPOSITION 1.3 : For a Banach space X, the following are equivalent

1. Relatively weakly compact sets in X are relatively norm compact.
2. Weakly convergent sequences in X are norm convergent.
3. WCC subsets of X are relatively norm compact.
4. Weak Cauchy sequences in X are norm convergent.
5. A bounded subset of X is either relatively norm compact or contains an ℓ^1-basis.
6. If Y is a Banach space and $T : Y \to X$ an operator, then T is either compact or fixes a copy of ℓ^1.

DEFINITION 1.4 : If X fulfils the above properties, then we say that X has the Schur-property.

Remark that a subspace of a Banach space with the Schur property also has the Schur property. The best known example of a Schur space is ℓ^1 itself. In fact, we have the following result, which is immediate from 1.3.

COROLLARY 1.5 : If X has the Schur property, then X is hereditarily ℓ^1.
The converse however is false.

The next result is in the same spirit as 1.3.

PROPOSITION 1.6 : For a Banach space X, the following are equivalent
1. Every WCC subset of X is relatively weakly compact.
2. Weak Cauchy sequences in X are weakly convergent.
3. A bounded subset of X is either relatively weakly compact or contains an ℓ^1-basis.
4. If Y is a Banach space and T : Y → X an operator, then T is either weakly compact or fixes a copy of ℓ^1.

DEFINITION 1.7 : If X satisfies the conditions of 1.6, we say that X is weakly sequentially complete (WSC).

Remark that again a subspace of a WSC space is WSC. Reflexive spaces are obviously WSC. All $L^1(\mu)$-spaces are WSC (see [50], IV).

1.6 has the following 2 immediate corollaries, without converse (cfr. [19]).

COROLLARY 1.8 : If X is Schur, then X is WSC.

COROLLARY 1.9 : A non-reflexive WSC space contains ℓ^1. Hence if X is WSC and has no infinite dimensional reflexive subspaces, then X is hereditarily ℓ^1.

DEFINITION 1.10 : We will say that a Banach space X has the Dunford-Pettis property (D-P), provided $\lim_n x_n^*(x_n) = 0$ whenever (x_n) is weakly null in X and (x_n^*) weakly null in X^*.
In fact, we may clearly assume one of the sequences (x_n), (x_n^*) only weakly Cauchy. So the D-P property means that weakly null sequences in X (resp. X^*) converge uniformely to 0 on WCC subsets of X^* (resp. X).

It follows immediately from the definition that X is D-P if X^* is D-P. The converse is false.

PROPOSITION 1.11 : A Banach space X is D-P iff the following holds If Y is a Banach space and $T : X \to Y$ a weakly compact operator, then T maps weak compact sets onto norm compact sets.

Proof : We have to show that if X is D-P and $T : X \to Y$ weakly compact, then $\lim_n \|Tx_n\| = 0$ whenever $\lim_n x_n = 0$ weakly in X. Suppose not, then there exists some $\varepsilon > 0$ and a sequence (y_n^*) in Y^* so that $\|y_n^*\| \leq 1$ and $\langle Tx_n, y_n^* \rangle > \varepsilon$ for each n (replacing (x_n) by a subsequence). Since also $T^* : Y^* \to X^*$ is weakly compact, $(T^* y_n^*)$ is relatively weakly compact in X^*. The fact that $\langle x_n, T^* y_n^* \rangle > \varepsilon$ for all n gives the contradiction.
Let us now pass to the converse and assume $\lim_n x_n = 0$ weakly in X and $\lim_n x_n^* = 0$ weakly in X^*. Consider the operator $T : X \to c_0$ mapping x onto $(\langle x, x_n^* \rangle)$. It is easily seen by dualisation that T is a weak compact operator. Thus, by hypothesis, $\lim_n \|Tx_n\| = 0$ and consequently also $\lim_n \langle x_n, x_n^* \rangle = 0$.

The next 2 corollaries are direct consequences of 1.11.

COROLLARY 1.12 : A reflexive Banach space is D-P if and only if it is finite dimensional.

COROLLARY 1.13 : Any complemented subspace of a D-P space is a D-P space.

It was shown by Grothendieck (see [63]) that $L^1(\mu)$ and $C(K)$ spaces are D-P spaces. The fact that ℓ^2 imbeds in L^1 shows that a subspace of a D-P space is not necessarily a D-P space.
Obviously Schur spaces are D-P.

2. ULTRAPRODUCTS AND $\mathcal{L}^p$-SPACES

Especially in the local Banach space theory, the notion of Banach-Mazur distance is important.

DEFINITION 1.14 : For normed linear spaces E and F, we let

$$d(E,F) = \inf \{\|T\| \, \|T^{-1}\| \ ; \ T : E \to F \text{ is an onto isomorphism}\}$$

(in case E and F are not isomorphic, take $d(E,F) = \infty$).

DEFINITION 1.15 : Assume X and Y Banach spaces. We say that X is finitely representable in Y provided for any finite dimensional subspace E of X and $\varepsilon > 0$, there is a subspace F of Y satisfying $d(E,F) < 1 + \varepsilon$.

We will now recall the definition of ultraproduct of Banach spaces and some related basic properties. This notion was introduced by D. Dacunha Castelle and J.L. Krivine in [82] and later studied more intensively by J. Stern [115] and S. Heinrich [65]

DEFINITION 1.16 : Let I be a set, $(X_i)_{i\in I}$ a family of Banach spaces and $\mathcal{U}$ a free ultrafilter on I.
The ultraproduct $(X_i)_{\mathcal{U}}$ of the (X_i) with respect to $\mathcal{U}$ will be the quotient space of the space

$\ell^\infty(X_i\ ;\ i \in I) = \{(x_i)\ ;\ x_i \in X_i \text{ and } \|(x_i)\| = \sup_i \|x_i\| < \infty\}$

by its subspace

$$N_{\mathcal{U}} = \{(x_i)\ ;\ \lim_{\mathcal{U}} \|x_i\| = 0\}$$

Let us remark that $\|(x_i)\|_{\mathcal{U}} = \lim_{\mathcal{U}} \|x_i\|$ for (x_i) in $(X_i)_{\mathcal{U}}$.

If the X_i are a same space X, then there is an obvious inbedding of X in $(X_i = X)_{\mathcal{U}} = (X)_{\mathcal{U}}$.

PROPOSITION 1.17 : The following classes of Banach spaces are stable under ultraproducts

1. Banach algebras
2. Banach lattices
3. C(K) spaces, for K compact Hausdorff
4. $L^p(\mu)$ spaces, where μ is a σ-additive measure and $1 \leq p < \infty$.

The proof for (3) and (4) relies on the abstract lattice characterization due to Kakutani [79], [80] for C(K) and $L^p(\ )$ spaces. The next three propositions are due to J. Stern [115].

PROPOSITION 1.18 : Any ultraproduct $(X)_{\mathcal{U}}$ of a space X is finitely representable in X.

PROPOSITION 1.19 :

1. Let $\mathcal{B}$ be a family of Banach spaces, X a Banach space and $c < \infty$ a constant. Suppose that for any finite dimensional subspace E of X and $\varepsilon > 0$ there is some $B \in \mathcal{B}$ and a subspace F of B so that $d(E,F) < c + \varepsilon$. Then there is an ultraproduct Y of elements of $\mathcal{B}$ and a subspace X' of Y such that $d(X,X') \leq c$.

2. If in particular X is finitely representable in Y, then X is isometric to a subspace of some ultraproduct $(Y)_{\mathcal{U}}$ of Y.

PROPOSITION 1.20 :

1. Let $\mathcal{B}$ be a family of Banach spaces, X a Banach space and $c < \infty$ a constant. Suppose that for any finite dimensional subspace E of X there exist some subspace F of X and some $B \in \mathcal{B}$ so that $E \subset F$ and $d(F,B) \leq c$. Then the bidual X^{**} of X is isomorphic to a complemented subspace of an ultraproduct of elements of $\mathcal{B}$.

2. In particular, X^{**} is isometric to a 1-complemented subspace of some ultraproduct $(X)_{\mathcal{U}}$ of X.

The proof of 1.18 and 1.19 is straightforward. The last result is a reformulation of the well-known principle of local reflexivity. We now pass to the definition of $\mathcal{L}^p$-spaces.

For $1 \leq p \leq \infty$ and $n = 1,2,\ldots$, we denote $\ell^p(n)$ the space $\mathbb{R}^n$ equipped with the ℓ^p-norm.

DEFINITION 1.21 : Assume $1 \leq p \leq \infty$ and $1 \leq \lambda < \infty$. A Banach space X is called a $\mathcal{L}^p_\lambda$-space provided for any finite dimensional subspace E of X, there is a finite dimensional subspace F of X satisfying $E \subset F$ and $d(F,\ell^p(n)) \leq \lambda$, where $n = \dim F$.
We say that X is $\mathcal{L}^p_{\lambda+}$ if X is $\mathcal{L}^p_{\lambda'}$ for all $\lambda' > \lambda$.

A $\mathcal{L}^p$ space is a $\mathcal{L}^p_\lambda$-space for some $\lambda < \infty$.

Thus $\mathcal{L}^p$ spaces are like ℓ^p from the local point of view. As we will see, this local property has however several implications on the global structure of these spaces.

PROPOSITION 1.22 : Any ultraproduct of $\mathcal{L}^p_\lambda$-spaces is a $\mathcal{L}^p_{\lambda+}$-space.

Proof : This follows easily from the definition and the fact that given a positive integer m and $\varepsilon > 0$, there exists some positive integer $n = n(m,\varepsilon)$ such that any m-dimensional subspace E of ℓ^p is contained in an n-dimensional subspace F of ℓ^p with $d(F,\ell^p(n)) < 1 + \varepsilon$.

PROPOSITION 1.23 :

1. If $1 < p < \infty$ and X is $\mathcal{L}^p$, then X is isomorphic to a complemented subspace of some $L^p(\mu)$-space.

2. If X is $\mathcal{L}^1$ (resp. $\mathcal{L}^\infty$), then X^{**} is isomorphic to a complemented subspace of an $L^1(\mu)$ (resp. $C(K)$)-space.

Proof : We will use 1.20 (1), taking for $\mathcal{B}$ the class of L^p-spaces if $1 \leq p < \infty$ and $C(K)$-spaces if $p = \infty$. The definition of $\mathcal{L}^p$-space yields us precisely the required condition. Consequently, taking also 1.17 in account, X^{**} is isomorphic to a complemented subspace of some element of $\mathcal{B}$. It follows that X is reflexive in case $1 < p < \infty$ and hence itself isomorphic to a complemented subspace of some $L^p(\mu)$.

It is well-known that for $1 \leq p < \infty$ any subspace of an $L^p(\mu)$-space in which ℓ^p is finitely representable, has a further subspace isomorphic to ℓ^p and complemented in this $L^p(\mu)$-space. Therefore the following result holds

PROPOSITION 1.24 : Any infinite dimensional $\mathcal{L}^p$-space $(1 \leq p < \infty)$ has a complemented subspace isomorphic to ℓ^p.

As will be shown later, 1.24 does not extend to the case $p = \infty$.

PROPOSITION 1.25 : If X is a $\mathcal{L}^p$-space and Y a complemented subspace of X containing ℓ^p (c_0 in case $p = \infty$), then Y is also $\mathcal{L}^p$.

Proof : Take for convenience $S = \ell^p$ if $1 \leq p < \infty$ and $S = c_0$ if $p = \infty$. Let X be a $\mathcal{L}^p_\lambda$-space. Denote $P : X \to Y$ the projection and let Z be a subspace of Y isomorphic to S.

Fix a finite dimensional subspace E of Y. Then E is contained in some subspace F of X which is λ-isomorphic to a finite dimensional ℓ^p-space. Denote $V = P(F)$ and $W = (I-P)(F)$. Because V is finite dimensional, it is possible to choose a subspace Z_1 of Z satisfying $d(Z_1,S) \leq d(Z,S)$ and $\max(\|y\|,\|z\|) \leq 4\|y+z\|$ if $y \in V$ and $z \in Z_1$. There is a subspace W' of Z_1 and an isomorphism $\iota : W \to W'$, where $\|\iota\|\|\iota^{-1}\|$ only depends on λ and $d(Z,S)$. Assume $\|\iota\| \leq 1$.
Introduce now the map $T : F \to Y$ defined by $Tx = Px + \iota(x-Px)$ and take $F' = T(F)$. Then $\|Tx\| \geq \frac{1}{4}\max(\|Px\|, \|\iota(x-Px)\|) \geq \frac{\|x\|}{8\|\iota^{-1}\|}$ for $x \in F$ and thus $d(F,F') \leq 24\|\iota^{-1}\|\|P\|$, as an easy computation shows. Moreover $E \subset F'$, since T is the identity on E. It follows that Y is a $\mathcal{L}^p_\mu$-space, where μ only depends on λ, $\|P\|$ and $d(Z,S)$.

The following results are well-known facts of classical Banach space theory : C(K)-spaces have the Pelczynski-property, which means that if X is a C(K)-space, Y a Banach space and $T : X \to Y$ an operator, then T is either weakly compact or fixes a copy of the space c_0. Any non-weakly compact operator T between L^1-spaces X and Y fixes a copy of the space ℓ^1.

Since complemented subspaces of C(K) and $L^1(\mu)$-spaces are D-P, we deduce from 1.12, 1.25 and the preceding

<u>PROPOSITION 1.26</u> : Complemented subspaces of C(K) (resp. $L^1(\mu)$) -spaces are $\mathcal{L}^\infty$ (resp. $\mathcal{L}^1$)-spaces.

The following dualization property holds

<u>PROPOSITION 1.27</u> : A Banach space X is $\mathcal{L}^p$ if and only if the dual X^* of X is $\mathcal{L}^q$, where p,q are as usually related by $p^{-1} + q^{-1} = 1$ $(\infty^{-1} = 0)$.

<u>Proof</u> : If X^{**} is $\mathcal{L}^p$, then X is also $\mathcal{L}^p$ by local reflexivity.
If $1 < p < \infty$ and X is an infinite dimensional $\mathcal{L}^p$-space, then X is isomorphic to a complemented subspace of $L^p(\mu)$ and has a complemented ℓ^p-subspace, by 1.23 and 1.24. Consequently X^* is isomorphic to a complemented subspace of $L^q(\mu)$ and has an ℓ^q-subspace. Applying 1.25, it follows that X^* is $\mathcal{L}^q$.

If X is $\mathcal{L}^1$ (resp. $\mathcal{L}^\infty$), then X^{**} is isomorphic to a complemented subspace of an $L^1(\mu)$ (resp. $C(K)$) space. Thus X^{***} and consequently X^* are isomorphic to a complemented subspace of a $C(K)$ (resp. $L^1(\mu)$) space and it remains to apply 1.26.

PROPOSITION 1.28 : A complemented subspace of a $\mathcal{L}^1$ (resp. $\mathcal{L}^\infty$) space is a $\mathcal{L}^1$ (resp. $\mathcal{L}^\infty$)-space.

Proof : If X is complemented in a $\mathcal{L}^1$ (resp. $\mathcal{L}^\infty$), then X^{**} is isomorphic to a complemented subspace of an $L^1(\mu)$ (resp. $C(K)$) space and hence $\mathcal{L}^1$ (resp. $\mathcal{L}^\infty$) (by 1.23 and 1.26). Therefore X is also $\mathcal{L}^1$ (resp. $\mathcal{L}^\infty$).

Since $\mathcal{L}^1$-spaces are isomorphic to subspaces of $L^1(\mu)$-spaces, we find

COROLLARY 1.29 : Any $\mathcal{L}^1$-space and in particular duals of $\mathcal{L}^\infty$-spaces are WSC.

The even duals (of order ≥ 2) of a $\mathcal{L}^\infty$-space are isomorphic to complemented subspaces of $C(K)$-spaces. Hence

COROLLARY 1.30 : $\mathcal{L}^1$, $\mathcal{L}^\infty$-spaces and all their duals have the D-P property.

In the next section, we will introduce $\mathcal{L}^\infty$-spaces in another way. Further properties of separable $\mathcal{L}^\infty$-spaces will be given in chapter III. More details concerning $\mathcal{L}^p$-spaces ($1 < p < \infty$) are also presented in chapter V.

3. CHARACTERIZATION OF $\mathcal{L}^\infty$-SPACES BY EXTENSION PROPERTIES OF OPERATORS

Let us start by recalling the definition of an injective Banach space.

DEFINITION 1.31 : A Banach space X is called injective provided the following holds :

If Y, Z are Banach spaces such that Y is a subspace of Z and $T : Y \to X$ is a bounded linear operator, then there is an extension operator $\tilde{T} : Z \to X$ such that $\tilde{T}y = Ty$ for all $y \in Y$.
This gives the following diagram

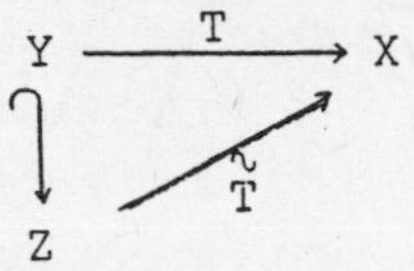

It is easy to see that if X is injective, some constant $\lambda > \infty$ can be found such that the operator $\tilde{T}$ can be choosen in such a way that $\|\tilde{T}\| \leq \lambda \|T\|$. In this case, X is called a P_λ-space.
Well-known examples of injective spaces are $\ell^\infty(\Gamma)$ (Γ being any set) and $L^\infty(\mu)$-spaces. In fact, any infinite dimensional injective space contains an isomorphic copy of ℓ^∞. For more details about injectivity, we refer the reader to [87] and H.P. Rosenthal's paper [108].

In the same spirit as 1.31, we may introduce extension properties with respect to certain classes of operators.

DEFINITION 1.32 : We say that a Banach space X has the compact (resp. weakly compact) extension property, if the following holds :
If Y, Z are Banach spaces such that Y is a subspace of Z and $T : Y \to X$ is a compact (resp. weakly compact) operator, then there exists a compact (resp. weakly compact) extension operator $\tilde{T} : Z \to X$.

The compact extension property was studied and completely characterized by J. Lindenstrauss [88]. The following result holds

PROPOSITION 1.33 : For a Banach space X, following properties are equivalent

1. X is a $\mathcal{L}^\infty$-space
2. X has the compact extension property
3. X^{**} is injective
4. X^{**} is a complemented subspace of a C(K)-space.

Let us give the idea of the proof. Since $\mathcal{L}^\infty$-spaces have a $\ell^\infty(n)$-local structure, finite rank operators into X can be extended with a bound on the norm of the extension operator. The extension of compact operators into X happens then by finite rank approximation.

If moreover X is isomorphic to a dual space, then a w^*-compactness argument allows us to extend any bounded operator into X and hence X is injective.
Suppose now X has compact extension property and denote K the closed unit ball of X^{***}. We see X^{**} as subspace of C(K), K endowed with the w^*-topology. By local reflexivity, it is possible to extend the identity on finite dimensional subspaces of X^{**} to bounded operators from C(K) into X^{**}. Again by compactness, this gives a projection of C(K) onto X^{**}.

Let us now pass to the weakly compact extension property.
We will use the following result, due to W.J. DAVIS, T. Figiel, W.B. Johnson and A. Pelczynski, for which the reader is referred to [40] or [43].

PROPOSITION 1.34 : If X is a Banach space and A a relatively weakly compact subset of X, then there exist a reflexive space Y and an operator $T : Y \to X$ such that A is contained in the image of the ball of Y.

The weakly compact extension property is characterized by

PROPOSITION 1.35 : The following properties for a Banach space X are equivalent
1. X has the weakly compact extension property
2. X is a $\mathcal{L}^\infty$-space possessing the Schur property

Proof :

(1) ⇒ (2) : By 1.33 and 1.34, it suffices to show that any operator $T : Y \to X$ with Y reflexive is compact. There exists a C(K)-space Z in which Y imbeds isometrically (take for instance K the closed unit ball of the dual Y^*). By hypothesis, T has a weakly compact extension $\tilde{T} : Z \to X$. Since the ball of Y is weakly compact and Z has the D-P property, $\tilde{T}$ maps the ball of Y onto a norm compact subset of X (1.11). Consequently T is compact.
(2) ⇒ (1) : Obvious by 1.33, since the compact and weakly compact operators into X are the same.

For a long time, one believed that infinite dimensional $\mathcal{L}^{\infty}$-spaces always have a c_0-subspace and consequently the only Banach spaces with weakly compact extension property are the finite dimensional spaces.
In chapter III, an infinite dimensional $\mathcal{L}^{\infty}$-space with the Schur property will be constructed.

REMARKS

1. It follows from the proof of 1.33 that in fact in the definition of compact extension property, we do not have to require the compactness of the extending operator $\tilde{T}$. This is obviously not the case for the weakly compact extension property.

2. There are other so called Hahn-Banach characterizations of $\mathcal{L}^{\infty}$-spaces, where now the space is the domain of the compact operator. These can be found in [85], [87] or [88].

II. GEOMETRICAL IMPLICATIONS OF CERTAIN FINITE DIMENSIONAL STRUCTURES

This chapter is in the first place an introduction to the next one. More precisely, we will discuss here Banach space structures which will occur in the particular examples constructed in the next chapter.
However, we think that the material presented here also has an independent interest, as will be clear from several other applications.

1. GEOMETRICAL INTRODUCTION

We say that a Banach space X has the Radon-Nikodym property (RNP) provided for every finite measure space (Ω,Σ,μ) and every μ-continuous vectormeasure $F : \Sigma \to X$ of finite variation, there exists a Bochner integrable function $f : \Omega \to X$ such that $F(E) = \int_E f\, d\mu$ for every $E \in \Sigma$.

The RNP of X is equivalent with the fact that all uniformely bounded X-valued martingales on a finite measure space are convergent a.e.
Besides these probabilistic characterizations of RNP, we can also introduce RNP as a geometrical property. Let us say that a subset A of X is dentable if for all $\varepsilon > 0$ there exists $x \in A$ satisfying $x \notin \overline{c}(A \setminus B(x,\varepsilon))$, where $B(x,\varepsilon)$ denotes the open ball with midpoint x and radius ε. If any (nonempty) bounded subset of X is dentable, then X is called a dentable Banach space. It is known that RNP and dentability of a space are equivalent properties. For the equivalence of those two (and also other) properties, we refer the reader to J. Diestel and J.J. Uhl's book [45].
In fact, a simple separation argument shows that dentability of a set A in a space X means that for any $\varepsilon > 0$ there exist some x^* in the dual X^* and $\alpha > 0$ so that the set

$$S(x^*,A,\alpha) = \{x \in A \; ; \; x^*(x) \geq \sup x^*(A) - \alpha\}$$

has diameter less than ε.

Following R. Phelps (see [100]), we will call the set introduced above a slice of A. It is now an easy exercice to see that a dentable Banach space X satisfies the property of existence of points of continuity.

(PC) : For any nonempty bounded and closed subset A of X, the identity on A has a weak-norm point of continuity.

Various authors pointed out that RNP is separably determined (cfr. [43], [45]). A similar argument shows that this is also true for (PC). More precisely, the following is true

THEOREM 2.1 : Suppose X is a space failing RNP (resp. PC). Then X has a subspace Y with a finite-dimensional Schauder decomposition failing RNP (resp. PC).

Recall that (P_n, M_n) is a finite dimensional Schauder decomposition (F.D.D) for the Banach space X iff each P_n is a continuous linear projection of X onto the finite dimensional M_n, $P_n P_m = 0$ if $n \neq m$ and $x = \Sigma_n P_n(x)$ for each $x \in X$.

Theorem 2.1 for the RNP follows from the following result proved in [20]. Assume X without RNP. Then for each $\lambda > 1$ there exist a subspace $\mathfrak{X}$ of X, a uniformely bounded $\mathfrak{X}$-valued martingale (ξ_n) and a sequence $(S_n : \mathfrak{X} \to \mathfrak{X})$ of finite rank projections, such that

1. $x = \lim_n S_n(x)$ for each $x \in \mathfrak{X}$
2. $\|S_n\| \leq \lambda$
3. $S_m S_n = S_n S_m = S_m$ if $m \leq n$
4. $S_n \xi_{n+1} = \xi_n$
5. (ξ_n) is nowhere convergent

The PC-case is easier and the argument is in fact also contained in [20].

In what follows, sufficient conditions on a decomposition will be given to ensure the RNP or PC property of the generated space.

2. SKIPPED BLOCKING PROPERTIES OF DECOMPOSITIONS

Assume a Banach space X given and a sequence (G_i) of subspaces of X.
Denote $[G_i]_i$ the closed linear subspace of X generated by the G_i.
For $1 \leqslant m \leqslant n \leqslant \infty$ we agree to take $G[m,n]$ the space $[G_i]_{i=m}^{n}$ generated by the G_i $(m \leqslant i \leqslant n)$.

The notion of P-skipped blocking property may be introduced in general as follows

DEFINITION 2.2 : Let P be a "class of finite dimensional decompositions". A sequence (G_i) of finite dimensional subspaces of a Banach space X is a P-skipped blocking decomposition (P-SBD) of X provided the following conditions are fulfilled

a) $X = [G_i]_{i=1}^{\infty}$

b) $G_i \cap [G_j]_{j \neq i} = \{0\}$ for all i

c) If (m_k) and (n_k) are sequences of positive integers so that $m_k < n_k + 1 < m_{k+1}$, then the sequence of spaces $G[m_k,n_k]$ is an FDD of class P for the subspace $[G[m_k,n_k]]_{k=1}^{\infty}$ of X.

X has the P-skipped blocking property (P-SBP) provided X possesses a P-skipped blocking decomposition.

Condition (b) means that for any i the space X is the direct sum of G_i and $[G_j]_{j \neq i}$. So we can introduce the projections P_i on the G_i, which clearly satisfy $P_i P_j = 0$ for $i \neq j$.
If P is "isomorphism invariant" then the P-SBP is clearly also preserved under isomorphism.
We will consider two classes of FDD's, namely the boundedly complete decompositions and the l^1-decompositions.
We recall that an FDD (M_n) of a space is called boundedly complete, if for any sequence (x_n), $x_n \in M_n$, the serie Σx_n converges provided the partial sums are bounded, an l^1-decomposition, provided there exists some $\delta > 0$ with $\|\Sigma_n x_n\| \geqslant \delta\, \Sigma_n \|x_n\|$; for any finite sequence (x_n) with $x_n \in M_n$.

REMARKS

1. It is easily seen that in the definition of l^1-SBP ; the constant δ in the l^1-decompositions may be choosen independent of the blocking.
2. In the definition of P-SBP we do not require that the sequence (G_i) is itself an F.D.D. of the space X, because this additional property is not really needed for our purpose. Let us also remark that any separable Banach space has the P-SBP if we don't make a restriction on the F.D.D.'s. The proof of this fact is standard (see lemma 2.4 below), by considering the given space as a subspace of a space with a basis.
3. In what follows, we are only interested in spaces with boundedly complete-SBP and l^1-SBP, but of course the above definition also makes sense for other types of decompositions.

PROPOSITION 2.3 : The boundedly complete-SBP and l^1-SBP are

1. Isomorphism invariant
2. Hereditary

The first claim is obvious. The second one is a direct consequence of the following lemma.

LEMMA 2.4 : Let S be an infinite dimensional linear subspace of the linear span of the G_i $(1 \leq i < \infty)$ in X. Then there exist sequences of positive integers (m_k) and (n_k) and finite dimensional subspaces H_k of S such that

1. $m_k < n_k + 1 < m_{k+2}$
2. S is generated by the H_k
3. $H_k \subset G[m_k, n_k]$
4. $H_k \cap [H_l]_{l \neq k} = \{0\}$

It is then indeed clear that (H_k) is a boundedly complete (resp. l^1)-SBD of the closure Y of S if the (G_i) are a boundedly (resp. l^1)-SBD of X.

Proof of the lemma 2.4 : Take a sequence (x_k) in S whose linear span equals S. Proceeding by induction on k, we introduce integers m_k, n_k, p_k and subspaces H_k, S_k of S so that the following properties hold

i. $m_k < p_k < n_k + 1 < p_{k+1}$

ii. $m_{k+1} = p_k$

iii. $x_k \in H_1 \oplus \dots \oplus H_k$

iv. $H_k \subset G[m_k, n_k]$

v. $S_k = G[p_k, \infty] \cap S$

vi. $S = H_1 \oplus \dots \oplus H_k \oplus S_k$

This construction is routine. We give the inductive step. It follows from (vi) that $x_{k+1} = y+z$ for some $y \in H_1 \oplus \dots \oplus H_k$ and $z \in S_k$. Now S_k is contained in the linear space generated by the G_i $(m_{k+1} \leqslant i < \infty)$ and therefore $z \in G[m_{k+1}, p_{k+1}-1]$ for some $p_{k+1} > n_k+1$. Let $S_{k+1} = G[p_{k+1}, \infty] \cap S$. Remark that $[z] + S_{k+1} = [z] \oplus S_{k+1}$. Because S_{k+1} has finite codimension in S_k, there is some finite dimensional subspace H_{k+1} of S_k satisfying $z \in H_{k+1}$ and $S_k = H_{k+1} \oplus S_{k+1}$. Take $n_{k+1} \geqslant p_{k+1}$ with $H_{k+1} \subset G[m_{k+1}, n_{k+1}]$. Clearly $x_{k+1} \in H_1 \oplus \dots \oplus H_k \oplus H_{k+1}$ and $S = H_1 \oplus \dots \oplus H_k \oplus H_{k+1} \oplus S_{k+1}$.
To complete the proof of the lemma, we have to verify that $H_k \cap [H_l]_{l \neq k} = \{0\}$ for all k. Now $[H_l]_{l \neq k} = H_1 \oplus \dots \oplus H_{k-1} \oplus [H_l]_{l>k} \subset H_1 \oplus \dots \oplus H_{k-1} \oplus G[p_k, \infty]$ and therefore $S \cap [H_l]_{l \neq k}$ $H_1 \oplus \dots \oplus H_{k-1} \oplus (S \cap G[p_k, \infty]) = H_1 \oplus \dots \oplus H_{k-1} \oplus S_k$. From (vi) $H_k \cap [H_l]_{l \neq k} = H_k \cap (H_1 \oplus \dots \oplus H_{k-1} \oplus S_k) = \{0\}$

A sequence (x_k) in X will be called ε-separated $(\varepsilon > 0)$ provided $\|x_k - x_l\| > \varepsilon$ for all $k \neq l$.

PROPOSITION 2.5 :

1. Any non relatively compact sequence in a Banach space X with boundedly complete-SBP has a difference-subsequence which is a boundedly complete basic sequence.
2. Assume X with l^1-SBP and let $\delta > 0$ be the decomposition constant (cfr. remark 1). Let (x_k) be a bounded sequence in X which is ε-separated. Then for $0 < \rho < \frac{1}{2}\varepsilon\delta$, there is a subsequence (y_k) of (x_k) which is a ρ-l^1-basis, i.e.

$$\|\Sigma_k a_k y_k\| \geqslant \rho \, \Sigma_k |a_k|$$

for all finite sequences (a_k) of scalars.

Proof : Again, this is a routine argument. For a given bounded sequence (x_k) in X and a sequence (ι_k) of positive numbers, it is always possible to find sequences of integers (m_k) and (n_k), a difference subsequence (v_k) of (x_k) and a sequence (w_k) in X such that :

i. $m_k < n_k+1 < m_{k+1}$

ii. $\|v_k-w_k\| < \iota_k$

iii. $w_k \in G[m_k,n_k]$

Now suppose (x_k) ε-separated and $\Sigma_k \iota_k = \iota < \varepsilon$. If the G_i are a boundedly complete-SBD, then of course (w_k) is a boundedly complete basic sequence and hence also $(v_k)_{k\geq k_0}$ for k_0 big enough, as a direct computation shows (cfr. [86], prop. 1.9).

Now if (G_i) is an l^1-SBD with constant δ, then (v_k) is a κ-l^1-basis, where $\kappa = \varepsilon\delta - (1+\delta)\iota$. Indeed, for scalars (a_k) we have

$$\begin{aligned}\|\Sigma_k a_k v_k\| &\geq \|\Sigma_k a_k w_k\| - \Sigma_k |a_k| \|v_k-w_k\| \\ &\geq \delta\, \Sigma_k |a_k| \|w_k\| - \Sigma_k |a_k| \iota_k \\ &\geq \varepsilon\delta\, \Sigma_k |a_k| - (1+\delta)\, \Sigma_k |a_k| \iota_k \\ &\geq \kappa\, \Sigma_k |a_k|\end{aligned}$$

Therefore, the given sequence (x_k) must have a ρ-l^1-subsequence for any $0 < \rho < \frac{\kappa}{2}$ (see [86], the proof of th. 2.e.5.).
This completes the proof.

COROLLARY 2.6 :

1. If X has boundedly complete-SBP and X^* is separable, then X is somewhat reflexive (any infinite dimensional subspace of X contains an infinite dimensional reflexive space).
2. A Banach space with l^1-SBP is a Schur space.

The first assertion follows from [72], prop. 3 and the second one is obvious.

REMARKS

4. We say that a Banach space X has the strong-Schur property provided there exists $\delta > 0$ such that any ε-separated bounded

sequence in X has a subsequence which is an $\varepsilon.\delta$-l^1-basis. Several examples of spaces are known which are Schur but not strong-Schur (cfr. [70]). Prop. 2.5 (2) says that spaces with l^1-SBP are strong-Schur. The converse however is false (see [31] and also remarks 7 and 8 below).

5. In the next chapter, a construction technique is given of pathological $\mathcal{L}^\infty$ spaces. The underlying structure of these spaces turns out to be a skipped blocking property. More precisely, the first class of spaces have l^1-SBP and the second class boundedly complete-SBP.

3. GEOMETRICAL PROPERTIES OF SKIPPED BLOCKING PROPERTIES

We will prove the following result

Theorem 2.7 : A. Banach spaces with boundedly complete-SBP have property PC

B. Banach spaces with l^1-SBP are RNP

We start by proving (A). The next lemma is completely elementary, so we omit its proof (cfr. [20], prop. 1).

LEMMA 2.8 : If a Banach space fails PC, then there exist a nonempty bounded subset A and some $\varepsilon > 0$, so that diam $U > \varepsilon$ for any nonempty weakly open subset U of A (diam means norm-diameter).

Let us now pass to the proof of the first part of 2.7.

Proof of (A) : Let (G_i) be the boundedly complete SBD of X and denote $P_i : X \to G_i$ the projections. If X fails PC, then there exist a subset A of X and $\varepsilon > 0$ such as in lemma 2.8. Proceeding again by induction on k, we construct sequences of integers (m_k) and (n_k), a sequence (x_k) of elements of A and a sequence (w_k) in X, such that

i. $m_k < n_k + 1 < m_{k+1}$

ii. $\|x_{k+1}-x_k\| > \frac{\varepsilon}{2}$

iii. $\|x_k - x_{k-1} - w_k\| < 2^{-k}$ $(x_0 = 0)$

iv. $w_k \in G[m_k, n_k]$

Let $m_1 = 1$. Take $x_1 \in A$ and choose w_1 in the linear span of the G_i so that $\|x_1 - w_1\| < 2^{-1}$. Thus $w_1 \in G[m_1, n_1]$ for n_1 big enough. Now suppose x_k and n_k obtained. Since

$$U = \{y \in A \; ; \; \Sigma_{i=1}^{n_k+1} \; \|P_i(y - x_k)\| < 2^{-k-2}\}$$

is a relative weak neighborhood of x_k, some $x_{k+1} \in U$ can be found with $\|x_{k+1} - x_k\| > \frac{\varepsilon}{2}$. Take w'_{k+1} in the linear span of the G_i so that $\|x_{k+1} - x_k - w'_{k+1}\| < 2^{-k-2}$ and $\Sigma_{i=1}^{n_k+1} \; \|P_i(w'_{k+1})\| < 2^{-k-2}$.

If we let

$$w_{k+1} = w'_{k+1} - \Sigma_{i=1}^{n_k+1} \; P_i(w'_{k+1})$$

then $\|x_{k+1} - x_k - w_{k+1}\| < 2^{-k-1}$ and $w_{k+1} \in G[m_{k+1}, n_{k+1}]$ for $m_{k+1} = n_k + 2$ and n_{k+1} large enough.

Remark that by (ii) and (iii) $\underline{\lim} \; \|w_k\| \geq \frac{\varepsilon}{2}$ and hence $\Sigma_k \; w_k$ is not convergent. But on the other hand

$$\|\Sigma_{k=1}^{K} \; w_k\| \leq \|\Sigma_{k=1}^{K} \; (x_k - x_{k-1})\| + \Sigma_{k=1}^{K} \; \|x_k - x_{k-1} - w_k\|$$

$$< \|x_K\| + \Sigma_{k=1}^{K} \; 2^{-k}$$

$$\leq \sup_{x \in A} \|x\| + 1 < \infty$$

for each integer K.

This contradicts however the fact that $(G[m_k, n_k])_{k=1}^{\infty}$ is a boundedly complete decomposition.

The proof of part (B) of theorem 2.7 is somewhat less direct. We need some geometrical preliminaries, which can be found in [20] also.

<u>LEMMA 2.9</u> : Let A be a bounded subset of X and let S be a slice of A. Assume further $\&$ a finite subset of X^* and $\varepsilon > 0$. Then there exists a slice S' of A such that S' $\subset$ S and the oscillation $O(x^* | S') < \varepsilon$ for any $x^* \in \&$.

Proof : Let $S = S(f,A,\alpha)$ for some $f \in X^*$ and $\alpha > 0$. Denote $\tilde{A}$ the w^*-closure of A in X^{**} and consider an extreme point x^{**} of $\tilde{A}$ so that $x^{**}(f) > \sup f(A) - \frac{\alpha}{2}$ (this is possible by the Krein-Milman theorem). Take $2\iota = \min(\varepsilon,\alpha)$. If we let

$$U = \{y^{**} \in \tilde{A} \ ; \ |y^{**}(x^*) - x^{**}(x^*)| < \iota \text{ for all } x^* \in \mathcal{E} \cup \{f\}\}$$

then x^{**} is not in the w^*-closure of the set $c(\tilde{A} \setminus U)$. Therefore a separation argument yields us some $g \in X^*$ and $\beta > 0$ with $S(g,A,\beta) \subset S(g,\tilde{A},\beta) \subset U$. If $x \in S' = S(g,A,\beta)$, then $f(x) > x^{**}(f) - \iota > \sup f(A) - \alpha$ and hence $x \in S$.
We also have for $x,y \in S'$ and $x^* \in \mathcal{E}$:

$$|x^*(x) - x^*(y)| \leqslant |x^*(x) - x^{**}(x^*)| + |x^*(y) - x^{**}(x^*)| < 2\iota \leqslant \varepsilon$$

as required.

LEMMA 2.10 : Let C be a bounded and convex subset of X and let U be a nonempty relatively weakly open subset of C. Then there is a finite number $S_1,\dots,S_p$ of slices of C and positive scalars $\lambda_1,\dots,\lambda_p$ so that

1. $\Sigma\, \lambda_q = 1$
2. $\Sigma_q\, \lambda_q\, S_q \subset U$

Proof : Denote $\tilde{C}$ the w^*-closure of C in X^{**} and E the set of the extreme points of $\tilde{C}$. Take some $x \in U$. There is a $\sigma(X^{**},X^*)$-neighborhood V of 0 in X^{**} with $(x + 2V) \cap C \subset U$. Since now $x \in C \subset \tilde{c}(E)$, it is possible to find points $x_1^{**},\dots,x_p^{**}$ in E and positive scalars $\lambda_1,\dots,\lambda_p$ with $\Sigma\, \lambda_q = 1$ and $\Sigma\, \lambda_q\, x_q^{**} \in x + V$. Using the extremality of the x_q^{**}, we can find for each $q = 1,\dots,p$ a slice $\tilde{S}_q$ of $\tilde{C}$ so that $\tilde{S}_q \subset x_q^{**} + V$. Then $S_q = \tilde{S}_q \cap C$ is a slice of C and $\Sigma\, \lambda_q\, S_q \subset \Sigma\, \lambda_q(x_q^{**} + V) \cap C \subset (x + 2V) \cap C$ and thus contained in U.

Proof of (B) of theorem 2.7 : Let (G_i) be the l^1-SBD of X and let $0 < \delta \leqslant 1$ be the decomposition constant. It follows from (A) that X has property PC. We assume X not RNP and work towards a contradiction. If X is not a dentable space, then X has a bounded and convex subset C such that diam $S > \varepsilon$ $(\varepsilon > 0)$ for any slice S of C.

Because X is PC, the identity map C, weak-C, norm has a point of continuity and this gives a nonempty weakly open subset U of C with diam U < $\frac{\varepsilon\delta}{4}$. Applying lemma 2.10, we obtain slices $S_1,\dots,S_p$ of C satisfying $\Sigma\,\lambda_q\,S_q \subset U$, for some convex combination $\lambda_1,\dots,\lambda_p$. We will now make a same type of construction as in the proof of (A). More precisely, we will introduce integers $(m_q)_{1\leqslant q\leqslant p}$ and $(n_q)_{1\leqslant q\leqslant p}$, elements $(x_q)_{1\leqslant q\leqslant p}$, $(y_q)_{1\leqslant q\leqslant p}$ in C and vectors $(w_q)_{1\leqslant q\leqslant p}$ in X satisfying following conditions

i. $m_q < n_q + 1 < m_{q+1}$ for $q = 1,\dots,p-1$

ii. $x_q \in S_q$, $y_q \in S_q$

iii. $\|x_q - y_q\| > \varepsilon$

iv. $\|x_q - y_q - w_q\| < \frac{\varepsilon\delta}{4}$

v. $w_q \in G[m_q,n_q]$

This construction is based on lemma 2.9 and the fact that any slice of C has diameter bigger than ε. We give the inductive step. Assume n_q obtained. It is then possible using lemma 2.9 to find a slice S'_{q+1} of C so that $S'_{q+1} \subset S_{q+1}$ and $\Sigma_{i=1}^{n_q+1}\ \|P_i(x-y)\| < \frac{\varepsilon\delta}{8}$ for x,y in S'_{q+1} (P_i is again the projection on G_i).

Choose x_{q+1}, y_{q+1} in S'_{q+1} with $\|x_{q+1} - y_{q+1}\| > \varepsilon$ and let w'_{q+1} be in the linear span of the G_i so that

$$\|x_{q+1} - y_{q+1} - w'_{q+1}\| < \frac{\varepsilon\delta}{8} \text{ and } \Sigma_{i=1}^{n_q+1}\ \|P_i(w'_{q+1})\| < \frac{\varepsilon\delta}{8}$$

Take then

$$w_{q+1} = w'_{q+1} - \Sigma_{i=1}^{n_q+1}\ P_i(w'_{q+1})$$

belonging to $G[m_{q+1},n_{q+1}]$ for $m_{q+1} = n_q+2$ and n_{q+1} big enough.

Clearly $\|x_{q+1} - y_{q+1} - w_{q+1}\| < \frac{\varepsilon\delta}{4}$

Since $(G[m_q,n_q])_{q=1}^{p}$ is a skipped blocking, we get

$$\|\Sigma_q\,\lambda_q\,w_q\| \geqslant \delta\,\Sigma_q\,\lambda_q\,\|w_q\|$$

and thus

$$\|\Sigma_q\,\lambda_q\,x_q - \Sigma_q\,\lambda_q\,y_q\| \geqslant \delta\,\Sigma_q\,\lambda_q\,\|w_q\| - \Sigma_q\,\lambda_q\|x_q - y_q - w_q\|$$

$$\geqslant \delta\,\Sigma_q\,\lambda_q\|x_q - y_q\| - (1+\delta)\,\Sigma_q\,\lambda_q\,\|x_q - y_q - w_q\| > \varepsilon\delta - (1+\delta)\frac{\varepsilon\delta}{4} > \frac{\varepsilon\delta}{2}$$

However, this is impossible, since diam $\Sigma_q\,\lambda_q\,S_q < \frac{\varepsilon\delta}{4}$

REMARKS

6. The known separable Banach spaces possessing the Radon-Nikodým property have a boundedly complete SBD. However, the structure of general RNP-spaces is not yet understood. In particular, it is unknown if any RNP-space contains a boundedly complete basic sequence.

7. There exist Banach spaces with boundedly complete SBP and failing RNP. So RNP is not a consequence of PC. Let X be the James-Hagler tree-space (cfr. [66]). Then, using the terminology of [66], X^* has a subspace F (spanned by the coordinate functionals φ^*) which turns out to have a boundedly complete SBD and fail the Radon-Nikodým property. The reader will find the details in [32].

8. In [31], a subspace of $L^1[0,1]$ is constructed which fails the Radon-Nikodým property and has a unit ball relatively compact in L^0. In particular, this space is strong-Schur.

9. It can be shown that for subspaces of L^1 the RNP and PC-property are equivalent. Thus any subspace of L^1 with boundedly complete SBP has RNP.

10. It seems unknown whether or not a subspace of L^1 with l^1-SBP is always isomorphic to a subspace of l^1.

III. NEW CLASSES OF $\mathcal{L}^\infty$-SPACES

We will present in this chapter a new construction technique for $\mathcal{L}^\infty$-spaces. This new spaces are important for two reasons. First, they solve several basic problems in $\mathcal{L}^\infty$-theory. Secondly, because they answer various questions of general Banach space theory. The material of this chapter is mainly contained in [29].

1. SUMMARY

We shall construct two classes $\mathfrak{X}$ and $\mathcal{Y}$ of Banach spaces. Any space X of $\mathfrak{X}$ will satisfy following properties

a) X is an infinite dimensional separable $\mathcal{L}^\infty$ space.

b) X is a Radon-Nikodým space. Since, as will be shown later, an infinite dimensional $\mathcal{L}^\infty$ space cannot be imbedded isomorphically into a separable dual space, this example solves negatively the following conjecture of J. Uhl (see [45]). Is every separable Radon-Nikodým space isomorphic to a subspace of a separable dual space ?

c) X is a Schur space. This answers negatively a conjecture of J. Lindenstrauss who asked in [88] whether a space with the weakly compact extension property is necessarily finite dimensional (cfr. 1.35).
In [84], A. Pelczynski and J. Lindenstrauss and in [85], J. Lindenstrauss and H. Rosenthal asked whether every $\mathcal{L}^\infty$-space contains a subspace isomorphic to c_0. Our example disproves this conjecture.

d) X is weakly sequentially complete. Since X^* is a $\mathcal{L}^1$-space, X^* is also weakly sequentially complete. For a long time it was conjectured that a Banach space is reflexive if and only if both X and X^* are weakly sequentially complete.

For members Y of $\mathcal{Y}$, the following hold

a') The Banach space Y is a separable infinite dimensional $\mathcal{L}^\infty$-space.

b') The Banach space Y is a Radon-Nikodým space

c') Y is somewhat reflexive, i.e. every infinite dimensional subspace of Y contains an infinite dimensional reflexive subspace.

Since Y is a $\mathcal{L}^{\infty}$-space it also has the Dunford-Pettis property and hence there are Dunford-Pettis spaces which are somewhat reflexive. For some time, the Dunford-Pettis property was understood as a property opposite to reflexivity. Following these lines W. Davis asked whether a somewhat reflexive space could have the Dunford-Pettis property.

We also show that

d') $\mathcal{Y}$ has a continuum number of non-isomorphic members. Hence there exists a continuum number of separable $\mathcal{L}^{\infty}$-types.

Let us recall at this point the following result due to D. Lewis and C. Stegall (see [89]).

THEOREM 3.1 : Let X be an infinite dimensional separable $\mathcal{L}^{\infty}$-space. Then

1. X^* is isomorphic either to l^1 or to $M[0,1]$ (the space of Radon measures on $[0,1]$)
2. X^* is isomorphic to l^1 if and only if l^1 does not imbed in X.

Consequently, X^* is isomorphic to $M[0,1]$ for X in $\mathfrak{X}$ and Y^* is isomorphic to l^1 for Y in $\mathcal{Y}$. It is surprising that there exist isomorphic preduals of l^1 not containing c_0.

The construction given in the next section is the first where $\mathcal{L}^{\infty}$-spaces are built using isomorphic copies of $l^{\infty}(n)$. The known examples of $\mathcal{L}^{\infty}$-spaces where constructed using isometric copies of $l^{\infty}(n)$ and therefore they all are isomorphic to preduals of L^1.

2. THE BASIC CONSTRUCTION

We will describe a general construction technique for $\mathcal{L}^{\infty}$-spaces, leading to the classes $\mathfrak{X}$ and $\mathcal{Y}$ by specification of certain real parameters.

THEOREM 3.2 : Let a, b and λ be real numbers so that

i. $0 \leq a \leq 1$ and $0 \leq b \leq 1$

ii. $\lambda > 1$

iii. $a + 2b\lambda \leq \lambda$

Then there exist an increasing sequence (d_n) of positive integers, an increasing sequence (x_n) of finite dimensional subspaces of l^∞ and operators $j_n : E_n = l^\infty(d_n) \to X_n$, satisfying the following properties

1. $\|j_n\| \leq \lambda$
2. j_n is an isomorphism and $\pi_n j_n$ is the identity on E_n
   ($\pi_n : l^\infty \to E_n$ denotes the restriction to the d_n first coordinates)
3. For $x \in X_n$, the following holds
   $$\|x\| = \max \begin{cases} \|\pi_n(x)\| \\ a\|\pi_m(x)\| + b\|x - j_m\pi_m(x)\| \quad (m < n) \end{cases}$$

In the remainder of this section, a, b and λ will be fixed numbers fulfilling (i), (ii), (iii) of 3.2.

We will first show how the sequence (d_n) is obtained and construct a system of injections $i_{m,n} : E_m \to E_n$ $(m < n)$ satisfying the two conditions

(α) $\pi_m \circ i_{mn} = id_{E_m}$ for $m < n$

(β) $i_{mn} \circ i_{lm} - i_{ln}$ for $l < m < n$

We shall give the technical description of the inductive procedure and afterwards some details for small n.

Take $d_1 = 1$.
Suppose now d_m $(m \leq n)$ known and the i_{lm} $(l < m \leq n)$ constructed such that they satisfy (α) and (β).
For $m < n$; $1 \leq i \leq d_m$; $1 \leq j \leq d_n$; $\varepsilon' = \pm 1$; $\varepsilon'' = \pm 1$, we define the functional $f_{m,i,j,\varepsilon',\varepsilon''} \in E_n^*$ as follows

$$f_{m,i,j,\varepsilon',\varepsilon''}(x) = a\varepsilon' x_i + b\varepsilon'' (x - i_{mn}\, \pi_m(x))_j$$

The subscripts i and j are referring to the respective coordinates. Remark that the above definition makes sense, because $\pi_m(x) \in E_m$.
Consider then the set of functionals

$$F_n = \{f_{m,i,j,\varepsilon',\varepsilon''} \mid m < n \ ; \ 1 \leq i \leq d_m \ ; \ 1 \leq j \leq d_n \ ; \ \varepsilon' = \pm 1 \ ; \ \varepsilon'' = \pm 1\}$$

Let $d_{n+1} = d_n + \text{card}\,(F_n)$ and enumerate the elements of F_n $g_{d_n+1}, \ldots, g_{d_{n+1}}$.

The map $i_{n,n+1} : E_n \to E_{n+1}$ is now defined as

$$i_{n,n+1}(x) = (x_1, x_2, \ldots, x_{d_n}, g_{d_n+1}(x), g_{d_n+2}(x), \ldots, g_{d_{n+1}}(x))$$

We put $i_{m,n+1} = i_{n,n+1}\, i_{m,n}$ for $m < n$.

The properties (α) and (β) remain trivially verified.
The heart of the construction lies in the metric properties of the injections. Before studying these properties, we give details for $n = 1,2,3$.

(i) <u>$n = 1$</u> : No possible value of m and hence $d_2 = d_1 = 1$ and $i_{1,2} = id_{E_1}$.

(ii) <u>$n = 2$</u> : Possible value of $m = 1$, possible value of $i = 1$, possible value of $j = 1$. It follows that card $(F_2) = 4$.

$f_{1,1,1,1,1}(x) = a\, x_1 + b\, (x - \pi_1(x))_1 = a\, x_1$

$f_{1,1,1,1,-1}(x) = a\, x_1$

$f_{1,1,1,-1.1}(x) = -a\, x_1$

$f_{1,1,1,-1,-1}(x) = -a\, x_1$

We obtain $i_{2,3}(x) = (x_1,\ a\, x_1,\ a\, x_1,\ -a\, x_1,\ -a\, x_1)$.
The number $d_3 = 5$.

(iii) <u>$n = 3$</u> : Possible values of $m = 1,2$

(1) For $m = 1$, one value of i, 5 values of j, hence 20 functionals

(2) For $m = 2$, one value of i, 5 values of j, hence 20 functionals.

Thus $d_4 = d_3 + 40 = 45$.
As a typical element, let us evaluate $f_{1,1,4,\varepsilon',\varepsilon''}$:

$$\begin{aligned} f_{1,1,4,\varepsilon',\varepsilon''}(x) &= a\, \varepsilon'\, x_1 + b\, \varepsilon''\, (x - i_{1,3}(x_1))_4 \\ &= a\, \varepsilon'\, x_1 + b\, \varepsilon''\, (x_4 - (-a\, x_1)) \\ &= (a\, \varepsilon' + b\, a\, \varepsilon'')x_1 + b\, \varepsilon''\, x_4 . \end{aligned}$$

<u>LEMMA 3.3</u> : The $i_{m,n}$ $(m < n)$ satisfy

(1) $\| i_{m,n} \| \leq \lambda$

(2) $\pi_m\ i_{m,n} = id_{E_m}$

(3) $i_{mn}\ i_{lm} = i_{ln}\ (l < m < n)$

(4) For $x \in E_n$, we have

$$\|i_{n,n+1}(x)\| = \max \begin{cases} \|x\| \\ a\|\pi_m(x)\| + b\|x - i_{mn}\pi_m(x)\| \ (m < n) \end{cases}$$

<u>Proof</u> : (2) and (3) were already observed.

(4) For $m < n$, the elements $i,j,\varepsilon',\varepsilon''$ can be choosen such that

$$\varepsilon'\ x_i = \|\pi_m(x)\|$$

$$\varepsilon''\ (x - i_{mn}\ \pi_m(x))_j = \|x - i_{mn}\ \pi_m(x)\|$$

Hence

$$\|i_{n,n+1}(x)\| \geq f_{m,i,j,\varepsilon',\varepsilon''}(x) = a\|\pi_m(x)\| + b\|x - i_{m,n}\ \pi_m(x)\|$$

The reverse inequality follows readily from the definition of $i_{n,n+1}$.

(1) Since $\|i_{1,2}\| = 1$, we can proceed by induction. Suppose we already have $\|i_{l,m}\| \leq \lambda$ for all $l < m \leq n$.

It follows from (4) that

$$\|i_{n,n+1}(x)\| \leq \max \begin{cases} \|x\| \\ a\|x\| + b(\|x\| + \|i_{mn}\|\|x\|) \ (m < n) \end{cases}$$

and hence by induction hypothesis

$$\|i_{n,n+1}(x)\| \leq \max\ (\|x\|,\ (a + b(1+\lambda))\|x\|) \leq \lambda\|x\|$$

Consequently $\|i_{n,n+1}\| \leq \lambda$.

If now $m < n$, then $i_{m,n+1}(x) = i_{n,n+1}\ i_{m,n}(x)$ and thus again by (4)

$$\|i_{m,n+1}(x)\| = \max \begin{cases} \|i_{m,n}(x)\| \\ a\|\pi_l\ i_{m,n}(x)\| + b\|i_{m,n}(x) - i_{l,n}\pi_l i_{m,n}(x)\| \text{ for } l < n \end{cases}$$

It remains to evaluate the second expression for all $l < n$.

We will distinguish the case $l \leq m$ and the case $m < l < n$.

If $l \leq m$, then $\pi_l\ i_{m,n} = \pi_l$ and hence

$$a\|\pi_l\ i_{m,n}(x)\| + b\|i_{m,n}(x) - i_{l,n}\ \pi_l\ i_{m,n}(x)\| =$$

$$a\|\pi_l(x)\| + b\|i_{m,n}(x) - i_{l,n}\ \pi_l(x)\| \leq$$

$$a\|x\| + b(\|i_{m,n}\|\|x\| + \|i_{l,n}\|\|x\|) \leq (a + 2b\lambda)\|x\| \leq \lambda\|x\|$$

If $m < l < n$, then $\pi_l\, i_{mn} = \pi_l\, i_{ln}\, i_{ml} = i_{ml}$ and $i_{ln}\, \pi_l\, i_{mn} = i_{ln}\, i_{ml} = i_{mn}$. Therefore

$$a\|\pi_l\, i_{mn}(x)\| + b\|i_{mn}(x) - i_{ln}\, \pi_l\, i_{mn}(x)\| = a\|i_{ml}(x)\| \leq a\, \lambda\, \|x\|.$$

We conclude that $\|i_{m,n+1}\| \leq \lambda$, proving the lemma.

We will see the E_n as subspaces of l^∞, using the canonical imbedding $E_n \hookrightarrow l^\infty : x \mapsto (x,0,0,\ldots)$.
Fix now n and E_n. For each $k > n$, the function $i_{n,k} : E_n \to E_k$ is constructed as above and thus at each stage new coordinates are added to the existing ones. Moreover, for each $x \in E_n$ the elements $i_{n,k}(x)$, $k > n$ are uniformely bounded by $\lambda\|x\|$. Hence, the limit $\lim\limits_{k\to\infty} i_{n,k}(x)$ exists in the w^*-topology of l^∞. Let us put

$j_n(x) = \lim\limits_{k\to\infty} i_{n,k}(x)$. Thus $j_n : E_n \to l^\infty$ is a linear operator and

we define X_n as the image of j_n, i.e. $X_n = j_n(E_n)$.

The properties of the j_n and the X_n are summarized in the next lemma, which will also complete the proof of theorem 3.2.

<u>LEMMA 3.4</u> :

(5) $\|j_n\| \leq \lambda$

(6) $\pi_n\, j_n = Id_{E_n}$

(7) $j_m = j_n\, i_{m,n}$ for all $m < n$

(8) $X_m \subset X_n$ for $m < n$

(9) $d(E_n, X_n) \leq \lambda$

(10) For $x \in X_n$, the following holds

$$\|x\| = \max \begin{cases} \|\pi_n(x)\| \\ a\|\pi_m(x)\| + b\|x - j_m\, \pi_m(x)\| \quad (m < n) \end{cases}$$

<u>Proof</u> :

(5) Clear from the bound $\|i_{n,k}(x)\| \leq \lambda\|x\|$ if $x \in E_n$ and $k > n$

(6) For $x \in E_n$, we have $\pi_n\, j_n(x) = \lim\limits_k \pi_n\, i_{n,k}(x) = x$ by (2).

(7) For $x \in E_m$ and $n > m$, we obtain

$$j_n\, i_{mn}(x) = \lim_k i_{nk}\, i_{mn}(x) = \lim_k i_{m,k}(x) = j_m(x)$$

(8) $X_m = j_m(E_m) = j_n\, i_{m,n}(E_m) \subset j_n(E_n) = X_n$.

(9) Since $\pi_n\, j_n = Id_{E_n}$, we find the following estimate for the Banach-Mazur distance

$$d(E_n,X_n) \leqslant \|j_n\|\|\pi_n\| \leqslant \lambda$$

(10) For $x \in X_n$, we find $\|x\| = \|j_n\, \pi_n(x)\| = \lim_{k\to\infty} \|\pi_k(x)\|$.

For $k \geqslant n$, it follows from (4) of lemma 3.3 that

$$\|\pi_{k+1}(x)\| = \|i_{k,k+1}\, \pi_k(x)\| = \max \begin{cases} \|\pi_k(x)\| \\ a\|\pi_m(x)\| + b\|\pi_k(x)-i_{m,k}\, \pi_m(x)\| \quad (m < k) \end{cases}$$

Now $i_{m,k}\, \pi_m(x) = i_{m,k}\, i_{n,m}(x) = i_{n,k}(x) = \pi_k(x)$ if $n \leqslant m < k$.
Consequently

$$\|\pi_{k+1}(x)\| = \max \begin{cases} \|\pi_k(x)\| \\ a\|\pi_m(x)\| + b\|\pi_k(x)-i_{m,k}\, \pi_m(x)\| \quad (m < n) \end{cases}$$

By iteration and using the fact that $\|\pi_k(x) - i_{m,k}\, \pi_m(x)\|$
$\|\pi_{k+1}(x) - i_{m,k+1}\, \pi_m(x)\|$, we see that

$$\|\pi_{k+1}(x)\| = \max \begin{cases} \|\pi_n(x)\| \\ a\|\pi_m(x)\| + b\|\pi_k(x)-i_{mk}\, \pi_m(x)\| \quad (m < n) \end{cases}$$

Because $\|x - j_m\, \pi_m(x)\| = \lim_{k\to\infty} \|\pi_k(x) - i_{m,k}\, \pi_m(x)\|$, we get (10).

So (3.4) and (3.2) are shown.

Let us now put X equal to the closure of $\cup_n X_n$ in l^∞. In fact, the space X can also be seen as the direct limit of the system

$$E_1 \xrightarrow{i_{1,2}} E_2 \xrightarrow{i_{2,3}} E_3 \longrightarrow \dots E_n \xrightarrow{i_{n,n+1}} E_{n+1} \longrightarrow \dots$$

The estimates $\|i_{n,n+1} \circ i_{n-1,n} \circ \dots \circ i_{2,1}\| \leq \lambda$ or the estimates (9) give that X is a $\mathcal{L}^{\infty}_{\lambda+}$-space. For more details about this simple fact, the reader is referred to [87].

It is clear from the preceding that X only depends on the parameters a and b. We will therefore call the space constructed above $X_{a,b}$.

<u>COROLLARY 3.5</u> : For all $x \in X_{a,b}$ and all m, the inequalities

$$\|x\| \geq a\|\pi_m(x)\| + b\|x - j_m\, \pi_m(x)\|$$

and consequently

$$\|x\| \geq a\|\pi_m(x)\| + b \text{ dist } (x,X_m)$$

hold.

<u>Proof</u> : First, let $x \in \bigcup_n X_n$, i.e. suppose $x \in X_n$ for some n. For $m < n$, the inequality is precisely estimate (10). If $m \geq n$, then $x = j_m\, \pi_m(x)$ and hence the inequality is trivial. For $x \in X$, we proceed by a density argument.

As we will see, the space $X_{a,b}$ turns out to be "pathological" as soon as $a + b > 1$. More precisely, the following two classes will be distinguished

$$\mathfrak{X} = \{X_{1,\delta} \; ; \; 0 < \delta < \tfrac{1}{2}\}$$

and

$$\mathcal{Y} = \{X_{a,b} \; ; \; 0 < b < \tfrac{1}{2} < a < 1 \text{ and } a + b > 1\}$$

The members of both classes are $\mathcal{L}^{\infty}$-spaces. This follows from the fact that some $\lambda > 1$ can be found satisfying (i), (ii), (iii) of 3.2, as an easy calculation shows.
We will study the classes $\mathfrak{X}$ and $\mathcal{Y}$ in the two next sections.

<u>3. $\mathcal{L}^{\infty}$-SPACES WITH THE SCHUR PROPERTY</u>

The structure of the members of $\mathfrak{X}$ is made clear by the following

THEOREM 3.6 : For any $\delta > 0$, the space $X_{1,\delta}$ has l^1-skipped-blocking-property.

Consequently, in virtue of 2.6 and 2.7, we have

COROLLARY 3.7 : Any member of the class $\mathfrak{X}$ has the Schur property and the Radon-Nikodým property.

Let $\delta > 0$ be fixed and take $X = X_{1,\delta}$. We will first construct a system of subspaces of X and then show that it forms an l^1-SBD. The same notations as in the previous section will be used.

For each n, the operator $S_n = j_n \pi_n : X \to X_n$ is a projection and $\|S_n\| \leq \lambda$. Let $P_1 = S_1$, $P_n = S_n - S_{n-1}$ $(n > 1)$ and denote $M_n = P_n(X)$. Then (M_n) is a finite dimensional Schauder decomposition for X. Fix no a strictly decreasing sequence (ρ_k) of real numbers so that $\frac{1}{2} < \rho_k < 1$. Proceeding then by induction on k, it is easy to construct an increasing sequence (n_k) of positive integers, such that

$$\|\pi_{n_{k+1}}(x)\| \geq \frac{\rho_{k+1}}{\rho_k} \|x\| \quad (*)$$

holds whenever $x \in X_{n_k}$.

We introduce for each k the subspace $G_k = [M_n \; ; \; n_{k-1} < n \leq n_k]$ $(n_0=0)$ which give also a finite dimensional decomposition of X.

Suppose now $(k(i))$ and $(l(i))$ increasing sequences of positive integers, such that $k(i) < l(i)+1 < k(i+1)$. Let (y_i) be a sequence in X so that

$y_i \in [G_k \; ; \; k(i) \leq k \leq l(i)] = [M_n \; ; \; n_{k(i)-1} < n \leq n_{l(i)}]$

By induction on j, we will establish the following inequality

$$\|\Sigma_{i=1}^{j} y_i\| \geq \delta \, \rho_{l(j)} \, \Sigma_{i=1}^{j} \|y_i\|$$

For $j = 1$, the inequality is obvious.
It follows from 3.5 and the above assumptions that, taking $n = n_{l(j)+1}$,

$$\|\Sigma_{i=1}^{j+1} y_i\| \geq \|\pi_n(\Sigma_{i=1}^{j+1} y_i)\| + \delta\|\Sigma_{i=1}^{j+1} (y_i - j_n \pi_n(y_i))\|$$

$$= \|\pi_n(\Sigma_{i=1}^{j} y_i)\| + \delta\|y_{j+1}\|$$

Because $\Sigma_{i=1}^{j} y_i \in X_{n_{l(j)}}$, we obtain from (*)

$$\|\Sigma_{i=1}^{j+1} y_i\| \geq \frac{\rho_{l(j)+1}}{\rho_{l(j)}} \|\Sigma_{i=1}^{j} y_i\| + \delta\|y_{j+1}\|$$

$$\geq \frac{\rho_{l(j)+1}}{\rho_{l(j)}} \delta\, \rho_{l(j)}\, \Sigma_{i=1}^{j} \|y_i\| + \delta\|y_{j+1}\|$$

$$\geq \rho_{l(j+1)}\, \delta\, \Sigma_{i=1}^{j+1} \|y_i\|$$

as required.

It is an immediate consequence of the preceding that the G_k form an l^1-SBD for X.

REMARKS

1. If X is a $\mathcal{L}^{\infty}_{1+}$-space, then it turns out that X^* is isometric to an L^1-space. In this case X is called a predual of L^1. In [118], M. Zippin has shown that such a space contains a subspace isometric to c_0. In [84], [85], [87] and [88], it is conjectured that

(i) If a Banach space X has the weakly compact extension property then X is finite dimensional

(ii) Every $\mathcal{L}^{\infty}$-space contains a subspace isomorphic to c_0

(iii) Every $\mathcal{L}^{\infty}$-space is isomorphic to a predual of L^1

(iv) Every $\mathcal{L}^{\infty}$-space is isomorphic to a quotient of a C(K)-space.

Since c_0 fails the Schur property, it follows from 1.35 that (i) is the weakest of all four conjectures. Any Banach space of the class $\mathfrak{X}$ satisfies the hypothesis of 1.35 and hence has the weakly compact extension property. This shows that all four conjectures have negative solution.

2. Because X^* is isomorphic to M[0,1], any member X of $\mathfrak{X}$ has a weakly sequentially complete dual. Since X is Schur, X is also WSC (1.8). As far as we know this space is the first non-reflexive space such that both X and X^* are WSC.

3. The argument given in the next section will also give a somewhat more direct proof of the RNP for members of $\mathfrak{X}$.

4. SOMEWHAT REFLEXIVE $\mathcal{L}^\infty$-SPACES

In this section, we will study the spaces $X_{a,b}$ for $0 < b < \frac{1}{2} < a < 1$ and $a + b > 1$. Assume a and b satisfying the above conditions fixed. We continue to use the terminology of section 2 and section 3.

THEOREM 3.8 : $Y = X_{a,b}$ has boundedly complete SBP.

Fix $\varepsilon > 0$ so that $\gamma = (a+b)(1-\varepsilon)^2 > 1$. Again, we introduce inductively a sequence (n_k) such that

$$\|\pi_{n_{k+1}}(x)\| \geqslant (1-\varepsilon)\ \|x\| \qquad (*)$$

for all $x \in X_{n_k}$.

We now claim that the spaces $G_k = [M_n\ ;\ n_{k-1} < n \leqslant n_k]$ $(n_0 = 0)$ form a boundedly complete SBD for Y.
In order to prove this, let $(k(i))$ and $(l(i))$ be increasing sequences of positive integers with $k(i) < l(i)+1 < k(i+1)$ for each i.
Let then (y_i) be a sequence in Y such that
$y_i \in [G_k\ ;\ k(i) \leqslant k \leqslant l(i)] = [M_n\ ;\ n_{k(i)-1} < n \leqslant n_{l(i)}]$.

Take $z_r = \Sigma_{i=1}^{r} y_i$ and assume (z_r) a bounded sequence. We have to show that then (z_r) is norm-convergent, or what amounts to the same $\{z_r\ ;\ r\}$ is norm-precompact.
If $\{z_r\ ;\ r\}$ is not precompact, then there exists $\beta > 0$ satisfying

$$\overline{\lim_{r}}\ \|z_r - S_m(z_r)\| \geqslant \beta \qquad (**)$$

for all m.

For each $m < n$ and r, we find by application of 3.5
$\|z_r - S_m(z_r)\| = \|z_r - j_m\ \pi_m(z_r)\| \geqslant$

$$a\|\pi_n(z_r - j_m\pi_m(z_r))\| + b\|z_r - j_m\pi_m(z_r) - j_n\pi_n(z_r - j_m\pi_m(z_r))\|$$

$$= a\|\pi_n(z_r - j_m\pi_m(z_r))\| + b\|z_r - j_n\pi_n(z_r)\|.$$

Choose then $1 \leqslant s < r$ and take $n = n_{1(s)+1}$. For $m < n$, it follows

$$\|z_r - S_m(z_r)\| \geqslant a\|\pi_n(z_s - j_m\pi_m(z_s))\| + b\|z_r - j_n\pi_n(z_r)\|$$

$$\geqslant a(1-\varepsilon)\,\|z_s - S_m(z_s)\| + b\|z_r - S_n(z_r)\|$$

using the fact that $z_s - S_m(z_s) \in X_{n_{1(s)}}$.

Given m, take s so that $n = n_{1(s)+1} > m$ and $\|z_s - S_m(z_s)\| > \beta(1-\varepsilon)$. For this n, we may choose arbitrarily large r satisfying $\|z_r - S_n(z_r)\| > \beta(1-\varepsilon)$. But then

$$\|z_r - S_m(z_r)\| \geqslant a(1-\varepsilon)^2\beta + b(1-\varepsilon)\beta > \gamma\beta$$

and hence

$$\overline{\lim_r}\, \|z_r - S_m(z_r)\| \geqslant \gamma\beta.$$

Hence, (**) holds for each m with β replaced by $\gamma\cdot\beta$. Repeating this argument t times yields

$$\infty > (1+\lambda)\sup_r \|z_r\| \geqslant \gamma^t\cdot\beta.$$

This is however a contradiction since $\gamma > 1$.

So far, only the lower estimates on the Y-norm were used. The next step will be to show that l^1 does not imbed in Y and this will require upper estimates.

PROPOSITION 3.9 : Y has no subspace isomorphic to l^1.

Proof : Assume (e_r) a sequence in Y equivalent to the usual basis of l^1. By passing eventually to a subsequence, we may suppose that (e_r) is weak* converging in l^∞. Take now $y_r = e_{2r} - e_{2r-1}$. This sequence is still equivalent to the usual basis of l^1 and $\pi_m(y_r) \to 0$ for each m. By density of $\cup_n X_n$ in Y, we can replace

(y_r) by a sequence (z_r) such that $\pi_m(z_r) = 0$ for $m < r$ and $z \in X_{m_r}$ where $m_1 < m_2 < \ldots < m_r < \ldots$

At this point, an important simplification is obtained by making use of the R.C. James regularization principle for l^1-sequences. It says that any l^1-sequence in a Banach space has a block-subsequence which behaves almost isometrically like the usual l^1-basis (see [68] for details).

Fix $\varepsilon > 0$ with $4\varepsilon < 1-a$. The James result yields a block subsequence (w_r) of (z_r) such that

1. $\|w_r\| = 1$
2. $\pi_m(w_r) = 0$ for $m < r$
3. $w_r \in X_{m'_r}$ where $m'_1 < m'_2 < \ldots < m'_r < \ldots$
4. $\|\Sigma'_r\, a_r\, w_r\| \geq (1-\varepsilon)\, \Sigma'_r\, |a_r|$

Define $x_1 = w_1$ and $n_1 = m'_1$. Let now r_2 be such that $\pi_{n_1}(w_r) = 0$ for $r \geq r_2$ and let $x_2 = w_{r_2}$ and $n_2 = m'_{r_2}$. Take then r_3 such that $\pi_{n_2}(w_r) = 0$ for $r \geq r_3$. Define $x_3 = w_{r_3}$ and $n = m'_{r_3}$.

We now calculate the norm of $x = x_1 + x_2 + x_3 \in X_n$. Of course, by (4), we have $\|x\| \geq 3 - 3\varepsilon$. The upper bounds of $\|x\|$ are given by application of 3.4 (10). Thus

$$\|x\| = \max \begin{cases} \|\pi_n(x)\| = \|\pi_n(x_1+x_2+x_3)\| \\ a\|\pi_m(x)\| + b\|x-j_m\pi_m(x)\| = \\ a\|\pi_m(x_1+x_2+x_3)\| + b\|x_1+x_2+x_3-j_m\pi_m(x_1+x_2+x_3)\| \end{cases} \quad (m < n)$$

We will first estimate the second type quantities by considering the different cases for m.

(α) $\underline{m \leq n_1}$:
$$a\|\pi_m(x)\| + b\|x-j_m\pi_m(x)\| =$$
$$a\|\pi_m(x_1)\| + b\|x_1+x_2+x_3-j_m\pi_m(x_1)\| \leq$$
$$a\|\pi_m(x_1)\| + b\|x_1-j_m\pi_m(x_1)\| + b\|x_2+x_3\|$$
$$\|x_1\| + b\|x_2+x_3\| \leq 1 + 2b.$$

(β) $\underline{n_1 < m \leq n_2}$: $a\|\pi_m(x)\| + b\|x - j_m\pi_m(x)\| =$

$a\|\pi_m(x_1+x_2)\| + b\|x_1+x_2+x_3-j_m\pi_m(x_1+x_2)\|$

$\leq a\|\pi_m(x_1+x_2)\| + b\|x_1+x_2-j_m\pi_m(x_1+x_2)\| + b\|x_3\|$

$\leq \|x_1+x_2\| + b\|x_3\| \leq 2 + b$

(γ) $\underline{n_2 < m < n}$: Since then $x_1 + x_2 = j_m\pi_m(x_1+x_2)$, we get

$a\|\pi_m(x)\| + b\|x-j_m\pi_m(x)\| =$

$a\|\pi_m(x_1+x_2+x_3)\| + b\|x_3-j_m\pi_m(x_3)\| \leq$

$a\|\pi_m(x_1+x_2)\| + a\|\pi_m(x_3)\| + b\|x_3-j_m\pi_m(x_3)\|$

$\leq a\ \|x_1+x_2\| + \|x_3\| \leq 2a + 1.$

If the norm of x is at least 3 - 3ε then this norm cannot be attained in the first d_{m_2} coordinates since there $x_3 = 0$. Hence, in order to estimate $\|\pi_n(x)\|$, we have to give bounds for the coordinates of x. situated between $d_{m_2} + 1$ and d_n. These coordinates are by construction bounded by the quantities

$$\|\pi_n(x_3)\| + a\|\pi_m(x_1+x_2)\| + b\|\pi_n(x_1+x_2)-i_{mn}\pi_m(x_1+x_2)\|$$

where $m < n_2$,
and hence by

$$1 + \max\ \{a\|\pi_m(x_1+x_2)\| + b\|x_1+x_2-j_m\pi_m(x_1+x_2)\|\ ;\ m < n_2\}$$

Again, we distinguish two cases, namely $m \leq n_1$ and $n_1 < m < n_2$. The reader will easily verify that the respective bounds 2 + b and 2 + a are obtained.

Summarizing (α), (β), (γ) and the $\|\pi_n(x)\|$-estimates gives
$\|x\| \leq \max\ (1+2b,\ 2+b,\ 1+2a,\ 2+b,\ 2+a) \leq 2+a < 3-4\varepsilon$
by the choice of ε.
This inequality contradicts however $\|x\| \geq 3-3\varepsilon$ and thus proves (3.9).

<u>REMARK</u> : Proposition 3.9 does not say that Y does not contain subspaces isomorphic (up to 1+ε) with $l^1(n)$. It only shows that such spaces are not spanned by "blocks".

As a consequence of 3.1 (2), 3.8 and 2.6, we obtain successively

THEOREM 3.10 : For any member Y of $\mathcal{Y}$, Y^* is isomorphic to l^1.

THEOREM 3.11 : Any member Y of $\mathcal{Y}$ is somewhat reflexive.

REMARK : The D-P property was understood as a property opposite to reflexivity. From this reasoning W.B. Davis asked following questions

(1) : Suppose Z is a Banach space without an infinite dimensional reflexive subspace. Is Z a D-P space ?

(2) : Suppose Z is a somewhat reflexive Banach space. Need Z to fail the D-P property ?

Both questions have a negative answer. Question 2 has a negative answer because the Banach spaces Y of class $\mathcal{Y}$ are somewhat reflexive and have as $\mathcal{L}^\infty$-spaces also the D-P property. Let us also remark that if Z is a Banach space which has the D-P property and does not contain an l^1-subspace, then Z^* is a Schur space.

Question 1 also has a negative answer. The following example will make this clear. Consider a sequence (p_i) of real numbers such that $1 < p_i < 2$ and $\lim_{i\to\infty} p_i = 1$. Let $B = \oplus_{l^1} (l^{p_i})$ be the l^1-sum of the spaces l^{p_i}, which is a subspace of L^1. Denote $(e_{i,n})_{n=1,2,\ldots}$ the sequence of the unit vectors of l^{p_i}. We then consider the subspace Z of B spanned by the weakly null sequence (f_n), where $f_n = \Sigma_{i=1}^\infty 2^{-i} e_{i,n}$. It is not very difficult to see that Z is hereditarily l^1 but fails the D-P property.
Other counterexamples to question 1 are obtained by considering certain Orlicz-sequence spaces.

It follows from 2.7 and 3.8 that any member of $\mathcal{Y}$ has the (PC)-property. Using a more specific argument, we will prove that in fact.

THEOREM 3.12 : Any space Y in $\mathcal{Y}$ has the Radon-Nikodym property.

The argument given here will also provide an alternative proof of the RNP for the members of class $\mathcal{X}$.

We refer the reader to [45] for the following basic result on representability of vector measures.

PROPOSITION 3.13 : Let (Ω,Σ,μ) be a probability space, Y a Banach space and $F : \Sigma \to Y$ a μ-continuous vector measure of bounded variation. Then F has a density if and only if there is a sequence of vector measures (F_n) such that each F_n has finite dimensional range and the variation norm $|F-F_n|$ tends to 0 for $n \to \infty$.

Remark that if $Y = X_{a,b}$, then 3.5 implies for each m

$$\|F(A)\| \geqslant a\|\pi_m F(A)\| + b\|F(A) - j_m\pi_m F(A)\| \text{ whenever } A \in \Sigma$$

Taking the supremum over all partitions of Ω, we obtain following inequality for the variation norms

$$|F| \geqslant a|\pi_m F| + b|F - j_m\pi_m F|$$

Proof of 3.12 : Let $Y = X_{a,b}$ be a member of $\mathcal{Y}$ and $F : \Sigma \to Y$ a bounded variation vector measure. For fixed n, define
$G = F - j_n\pi_n F$.
As remarked above, we find for all m

$$|G| \geqslant a|\pi_m G| + b|G - j_m\pi_m G|$$

Let $A_1,\ldots,A_r$ be a partition of Ω such that

$$|G| \leqslant \Sigma_{s=1}^{r} \|G(A_s)\| + \frac{\varepsilon}{2}$$

and take m_0 large enough to ensure that

$$\Sigma_{s=1}^{r} \|\pi_m G(A_s)\| \geqslant \Sigma_{s=1}^{r} \|G(A_s)\| - \frac{\varepsilon}{2} \text{ for } m \geqslant m_0$$

It follows that $|G| \leqslant |\pi_m G| + \varepsilon$ and hence $|G-j_m\pi_m G| \leqslant \frac{1-a}{b}|G| + \frac{a\varepsilon}{b}$.

Now for $m \geqslant n$, we have $G - j_m\pi_m G = F - j_n\pi_n F - (j_m\pi_m F - j_n\pi_n F)$ $= F - j_m\pi_m F$. Hence for $m \geqslant \max(m_0,n)$

$$|F - j_m\pi_m F| \leqslant \frac{1-a}{b} |F - j_n\pi_n F| + \frac{a\varepsilon}{b}.$$

Thus also $\overline{\lim_m} |F - j_m\pi_m F| \leq \frac{1-a}{b} |F - j_n\pi_n F| + \frac{a\varepsilon}{b}$

and $\overline{\lim_m} |F - j_m\pi_m F| \leq \frac{1-a}{b} \overline{\lim_n} |F - j_n\pi_n F| + \frac{a\varepsilon}{b}.$

Since this inequality holds for all $\varepsilon > 0$ and $a + b > 1$, we obtain $\overline{\lim_n} |F - j_n\pi_n F| = 0$. Thus F is the limit in variation norm of the finite rank measures $j_n\pi_n F$ and by 3.13 this completes the proof.

5. THE UHL CONJECTURE ON SEPARABLE RNP-SPACES

The origin of this conjecture lies in the following nice result of C. Stegall (see [114]) on the structure of dual spaces.

THEOREM 3.14 : If X is a Banach space, then X^* has the RNP if and only if S^* is separable for all separable subspaces S of X.

For a long time, the only known examples of separable RNP spaces were the separable duals and their subspaces. It follows easily from 3.14 that any separable subspace of an RNP dual imbeds in a separable dual space. J.J. Uhl conjectured that every separable Radon-Nikodým space is isomorphic to a subspace of a separable dual (see [45], p. 82 and p. 211-212). Different geometric characterizations of the Radon-Nikodým property are known. If Uhl's conjecture were true then a beautiful geometric description of subspaces of separable dual spaces would be obtained.
Unfortunately, the conjecture is wrong. Indeed, the members of both classes $\mathfrak{X}$ and $\mathcal{Y}$ have the RNP and the following result holds

PROPOSITION 3.15 : Let X be an infinite dimensional $\mathcal{L}^\infty$ space. If Y is a Banach space such that X is isomorphic to a subspace of Y^*, then Y contains a complemented subspace Z which is isomorphic to l^1. In particular, Y^* is not separable.

Proof : Let $i : X \to Y^*$ be the inclusion map. Transposition gives a map : $T : Y \to X^*$ defined as $T(y)(x) = i(x)(y)$. It follows that $T^* : X^{**} \to Y^*$ and that $T^*|_X = i$.
If T is a weakly compact operator then also T^* is a weakly compact operator and hence the inclusion map i being a restriction of T^*

is also a weakly compact operator. Since X is an infinite dimensional $\mathcal{L}^\infty$ space it is not a reflexive space and hence i is not weakly compact. It follows that T is not weakly compact. Hence the image T(B(Y)) is not a relatively weakly compact subset of X^*.
We know that X^* is isomorphic to a complemented subspace of an L_1 space and hence the Kadec-Pełczynski theorem [78] applies. The set T(B(Y)) contains a sequence $(e_n)_{n\geq 1}$ such that $(e_n)_{n\geq 1}$ is equivalent to the usual basis of l_1 and such that $S = \overline{\text{span}}\ (e_n, n \geq 1)$ is complemented in X^*.
Let $P : X^* \to S$ be a continuous projection. Take for each $n \geq 1$ an element $y_n \in B(Y)$ such that $T(y_n) = e_n$. It is elementary to see that $(y_n)_{n\geq 1}$ is equivalent to the usual basis of l_1 and also to $(e_n)_{n\geq 1}$. Let now $Z = \overline{\text{span}}\ (y_n, n \geq 1)$ and let $V : S \to Z$ be the operator defined by the relation $V(y_n) = e_n$. If Q is defined as $Q = V \circ P \circ T$ then clearly Q is a projection $Y \to Z$.

REMARK : At about the same time the $\mathcal{L}^\infty$-spaces presented in this chapter were discovered, P. Mc Cartney and R.C. O'Brien (see [31]) obtained independently an example of a separable Radon-Nikodým space which does not imbed in a separable conjugate space. The main idea in their construction is a notion which they call "neighborly-tree-property". This property is an isomorphism invariant and duals with the neighborly-tree-property always fail the RNP (a w^*-compactness argument allows to replace the "neighborly-tree" by a real diadic tree).
On the other hand, it turns out that there are easy examples of Radon-Nikodým spaces which have neighborly-tree-property and hence cannot imbed in a Radon-Nikodým dual.
Our $\mathcal{L}^\infty$-spaces are counterexamples to the Uhl-conjecture for a different reason. It can be shown that the members of $\mathfrak{X}$ not only have RNP but also fail the neighborly-tree-property.

6. ON THE NUMBER OF SEPARABLE $\mathcal{L}^\infty$-TYPES

Of course, the best known separable $\mathcal{L}^\infty$-spaces are the C(K)-spaces for K compact metric. From the isomorphic point of view the structure of these spaces is described in the following two results.

PROPOSITION 3.16 : If K is an uncountable compact metric space, then C(K) is isomorphic to C(Δ), the space of continuous functions on the Cantor set.

PROPOSITION 3.17 : Suppose K and L countable compact spaces with respective Cantor-ordinals α and β. Then C(K) and C(L) are isomorphic Banach spaces iff $\alpha^{\omega} = \beta^{\omega}$.

(3.16) is due to Milutin (see [93]) and (3.17) to Bessaga and Pełczynski [13]. It is an immediate consequence of these results that there exist only $\aleph_1$ mutually non-isomorphic C(K)-spaces for K compact metric.
The aim of this section is to show that $\mathcal{Y}$ has a continuum number of mutually non-isomorphic members. This gives a continuum number of separable $\mathcal{L}^{\infty}$-types.
$\mathcal{Y}$ consists of the spaces $X_{a,b}$ where $0 < b < \frac{1}{2} < a < 1$ and $a + b > 1$. More precisely, it will be proved that for fixed a the Banach spaces $(X_{a,b})_b$ are mutually non-isomorphic.

To prove that the spaces $X_{a,b}$ are not isomorphic, we will construct a basic sequence (e_k) in $X_{a,b}$, such that the estimate

$$\|e_{k_1} + e_{k_2} + \ldots + e_{k_N}\| \leqslant N^{\alpha}$$

holds, whenever $k_1 < k_2 < \ldots < k_N$. Here α is the unique number such that

$$a^{\frac{1}{1-\alpha}} + b^{\frac{1}{1-\alpha}} = 1.$$

On the other hand we show that if $\|x_k\| = 1$ and $x_k \to 0$ weakly in $X_{a,b}$, then there is a subsequence (z_k) of (x_k) and a constant $c > 0$, such that $\|z_1 + z_2 + \ldots + z_N\| \geqslant c\, N^{\alpha}$ for each N.

Since for fixed a the parameter α is in one-one correspondence with b this shows that the spaces $X_{a,b}$ are not isomorphic. More precisely, $X_{a,b}$ is not isomorphic to a subspace of $X_{a,b'}$ if $b < b'$.

Our presentation here will be sketchy. The reader will find more details in [29].

LEMMA 3.18 : There is a sequence of natural numbers $(n_k)_{k\geq 1}$ with following properties

1. $d_3 < n_1 \leq d_4 < n_2 \leq d_5 \ldots \leq d_{k+2} < n_k \leq d_{k+3} \ldots$
2. If $x \in E_m$ and $x_1 = x_2 = 0$ then the n_k-coordinate of $j_m(x)$ is zero for all $k' > m$.

LEMMA 3.19 : If $x \in E_m$ and $\pi_{m-1}(x) = 0$, then $\|x\| = \|j_m(x)\|$.

(3.18) is immediate from the construction of the spaces $X_{a,b}$ and (3.19) follows from (3.4).

Let now e'_k be the element of E_{k+3} defined as $(e'_k)_i = 1$ if $i = n_k$ and $(e'_k)_i = 0$ if $i \neq n_k$. Put then $e_k = j_{k+3}(e'_k)$. From the preceding, we get $\|e_k\| = 1$.

From the estimates of lemma 3.4 and from (3.18) we deduce that the sequence (e_k) is an unconditional basis (of constant 1) and that for $k_1 < \ldots < k_N$, the norm of $x = e_{k_1} + \ldots + e_{k_N}$ is attained in the extension beyond the coordinate d_{k_N+3}. Hence

$$\|x\| = \max_{m<k_N+3} (a\|\pi_m(x)\| + b\|x-j_m\pi_m(x)\|)$$

$$\leq \max_{n\leq N} (a\|e_{k_1}+\ldots+e_{k_n}\| + b\|e_{k_{n+1}}+\ldots+e_{k_N}\|)$$

For each N, define $\gamma_N = \sup \{\|e_{k_1}+\ldots+e_{k_N}\| \ ; \ k_1 < \ldots < k_N\}$

The above estimate then gives (1) $\gamma_1 = 1$

$$(2) \quad \gamma_N = \max_{n\leq N} (a\gamma_n + b\gamma_{N-n})$$

Proceeding by induction on N, the following is easily obtained by elementary calculus.

LEMMA 3.20 : If (γ_N) is a sequence of real numbers satisfying (1) and (2), then

$$\gamma_N \leq N^\alpha$$

where α is the unique number between 0 and 1 satisfying

$$a^{\frac{1}{1-\alpha}} + b^{\frac{1}{1-\alpha}} = 1$$

Let us now pass to the reverse subsequence estimations.
Suppose (x_k) a sequence in $X_{a,b}$ such that $\|x_k\| = 1$ and $x_k \to 0$ weakly.
By a density argument, we may of course assume (x_k) in $\cup_n X_n$.
Take a decreasing sequence (ε_k) of positive numbers with $\Pi_k(1-\varepsilon_k)>0$.

It is possible to extract a subsequence (z_k) of (x_k) satisfying following conditions

1. $z_k \in X_{s_k}$

2. $\pi_{s_k}(z_l) = 0$ if $l > k$

3. $\|\pi_{s_k}(x)\| \geq (1-\varepsilon_k)\|x\|$ for all $x \in \text{span}\,(z_1,\ldots,z_k)$

for some sequence $s_1 < s_2 < \ldots < s_k < \ldots$ of natural numbers.

Put now $\delta_N = \inf_k \|z_{k+1} + \ldots + z_{k+N}\|$.

Clearly $\delta_1 = 1$ and by (3.5)

$$\|z_{k+1} + \ldots + z_{k+N}\| \geq$$

$$a\|\pi_{s_{k+n}}(z_{k+1}+\ldots+z_{k+N})\| + b\|z_{k+1}+\ldots+z_{k+N}-j_{s_{k+n}}\pi_{s_{k+n}}(z_{k+1}+\ldots+z_{k+N})\|$$

$$= a\|\pi_{s_{k+n}}(z_{k+1}+\ldots+z_{k+n})\| + b\|z_{k+n+1} + \ldots + z_{k+N}\| \geq$$

$$a(1-\varepsilon_n)\delta_n + b\,\delta_{N-n}.$$

Thus (1) $\delta_1 = 1$

(2) $\delta_N \geq \max_{n\leq N} (a(1-\varepsilon_n)\delta_n + b\,\delta_{N-n})$

The next lemma is similar but more technical than (3.20).

<u>LEMMA 3.21</u> : Under the above conditions (1) and (2), there exists a constant $c > 0$ such that $\delta_N \geq c\,N^{\alpha}$, where α is defined as in lemma 3.20.

The proof is left as an instructive exercice for the reader.

THEOREM 3.22 : If $b' > b$, then $X_{a,b}$ is not isomorphic to a subspace of $X_{a,b'}$.

Proof : If $b < b'$, then $\alpha < \alpha'$ where $a^{\frac{1}{1-\alpha}} + b^{\frac{1}{1-\alpha}} = 1$ and $a^{\frac{1}{1-\alpha'}} + b'^{\frac{1}{1-\alpha'}} = 1$. Let (e_n) be the sequence in $X_{a,b}$ constructed above. Since $\overline{\text{span}}\,(e_n \,;\, n \geq 1)$ does not contain a copy of l^1 and since (e_n) is unconditional we obtain that $e_n \to 0$ weakly. Suppose now $X_{a,b}$ isomorphic to a subspace Z of $X_{a,b'}$. Let x_n be the element of Z corresponding to e_n. Since $x_n \to 0$ we also have that $z_n = \frac{x_n}{\|x_n\|} \to 0$ weakly. By passing to a subsequence and by application of lemma 3.21, we obtain a constant $c > 0$ such that $\|z_1+\ldots+z_N\| \geq c\, N^{\alpha'}$.
Consequently $\|x_1 + \ldots + x_N\| \geq c'\, N^{\alpha'}$, for some $c' > 0$.
On the other hand, there is some constant $c'' < \infty$, such that $\|x_1 + \ldots + x_N\| \leq c''\|e_1 + \ldots + e_N\| \leq c''\, N^{\alpha}$.
This however is a contradiction.

7. RELATED REMARKS AND PROBLEMS

1. It is not clear whether or not the spaces $X_{1,b}$ are isomorphic for different values of b.

2. One may raise the question if the spaces $X_{a,b}$ are prime, i.e. if any complemented subspace of $X_{a,b}$ is finite dimensional or isomorphic to $X_{a,b}$.

3. It turns out to be difficult (and in fact we are unable) to modify the construction in order to obtain $\mathcal{L}^\infty$-spaces without c_0-subspace and failing the RNP.

4. One of the important questions in general Banach space theory is the following :
   Does any infinite dimensional Banach space contain c_0 or l^1 or a reflexive subspace ?
   Even for $\mathcal{L}^\infty$-spaces, the problem is open.

5. It is not known if any infinite dimensional Banach space contains an unconditional basic sequence. In fact, we don't know

if any subspace of $X_{a,b}$ $(a < 1)$ contains an unconditional basic sequence.

6. The technique described in this chapter is not only a new way of constructing $\mathcal{L}^{\infty}$-spaces but also Banach spaces in general. Moreover there are also clearly certain modifications possible which may be interesting to investigate.

IV. A CLASS OF $\mathcal{L}^p$-SPACES

In this chapter a new family of complemented subspaces of $L^p[0,1]$ $(1 < p < \infty)$ will be introduced and studied. These results are presented in [30] in a somewhat different way. For certain facts, the reader will also be referred to [87] and [60].

1. INTRODUCTION

Unless otherwise specified, we always suppose $1 < p < \infty$.
It was already pointed out in 1.23 that any $\mathcal{L}^p$-space is isomorphic to a complemented subspace of an $L^p(\mu)$-space. Conversely, a complemented subspace of $L^p(\mu)$ containing a copy of l^p is a $\mathcal{L}^p$-space (see 1.25). In fact, the following holds

PROPOSITION 4.1 : If X is a complemented subspace of $L^p(\mu)$, then X is either a $\mathcal{L}^p$-space or isomorphic to a Hilbert space.

Proof : By duality, only the case $2 \leq p < \infty$ must be considered.
Define for fixed $\eta > 0$ and $f \in X$ the quantity
$\iota_\eta(f) = \mu[\,|f| \geq \eta\|f\|_p]$

The following two cases are distinguished
(a) $\inf\{\iota_\eta(f)\ ;\ f \in X\} = 0$ for any $\eta > 0$
(b) There exists $\eta > 0$ and $\rho > 0$ such that $\iota_\eta(f) > \rho$ whenever $f \in X$

It is easily seen that in case (a) the space l^p embeds in X and therefore X is a $\mathcal{L}^p$-space.
If (b), then it is clear that for all $f \in X$

$$\|f\|_2 \leq \|f\|_p \leq \frac{1}{\eta\sqrt{\rho}}\|f\|_2$$

holds. Consequently, X is isomorphic to a subspace of $L^2(\mu)$ and thus a Hilbert space.

On the other hand, we have

PROPOSITION 4.2 : Suppose $2 \leq p < \infty$ and X a subspace of $L^p(\mu)$ isomorphic to a Hilbert space. Then X is complemented in $L^p(\mu)$.

Proof : The case $p = 2$ is trivial. So let $2 < p < \infty$. Since X has no l^p-subspace, we are necessarily in alternative (b) of (4.1). Thus the $\| \|_p$ and the $\| \|_2$ norm coincide on X. Therefore, the ortogonal projection P on X is bounded in $\| \|_p$-norm, since

$$\|P(f)\|_p \sim \|P(f)\|_2 \leqslant \|f\|_2 \leqslant \|f\|_p$$

for all $f \in L^p(\mu)$.

Let us recall the classical Khintchine inequalities for the Rademacker functions (r_n) on $[0,1]$.

PROPOSITION 4.3 : For any $1 < p < \infty$, there is a constant $K_p < \infty$ such that

$$\frac{1}{K_p} (\Sigma_n |a_n|^2)^{1/2} \leqslant \|\Sigma_n a_n r_n\|_p \leqslant K_p(\Sigma_n |a_n|^2)^{1/2}$$

for any finitely supported sequence (a_n) of scalars.

For $p = 2$, we obviously find $K_2 = 1$.
(4.3) is a particular case of a more general result which we will present later.
Thus the Rademacker functions span a Hilbertian subspace of $L^p[0,1]$ for all $1 < p < \infty$.
It follows then from 4.2 and a dualization argument that $[r_n ; n]$ is complemented in $L^p[0,1]$ for each $1 < p < \infty$. Consequently $L^p[0,1]$ has complemented Hilbertian subspaces for each $1 < p < \infty$. Let us remark at this point that L^p also has non-complemented Hilbertian subspaces in the case $1 < p < 2$ (see [57]).

As an immediate consequence of (1.25) and the preceding, we get

PROPOSITION 4.4 : The spaces $Y_p = l^2 \oplus l^p$ and $Z_p = (l^2 \oplus l^2 \oplus \ldots)_p$, i.e. the l^p-sum of l^2-spaces, are $\mathcal{L}^p$-spaces.

For some time, the spaces l^p, L^p, Y_p and Z_p were the only known $\mathcal{L}^p$-spaces.
The next step was H. Rosenthal's discovery of the space X_p.

Assume $2 \leq p < \infty$. Let $\omega = (\omega_n)$ be a sequence of real numbers with $0 \leq \omega_n \leq 1$ for all n. We then consider the space $X_{p,\omega}$ consisting of all sequences of reals $x = (x_n)$, so that

$$\|x\| = \max \{(\Sigma |x_n|^p)^{1/p}, (\Sigma \omega_n^2 |x_n|^2)^{1/2}\} < \infty$$

Thus $X_{p,\omega}$ is a subspace of Y_p. The interest of these spaces lies in their probabilistic interpretation (see [109] and [110]).

PROPOSITION 4.5 : Assume $2 \leq p < \infty$ and (f_n) a sequence of independent random variables on $[0,1]$ with $\|f_n\|_p = 1$ and $\int_0^1 f_n(t)\,dt = 0$ for each n. Then (f_n) is equivalent to the unit vector basis of $X_{p,\omega}$ where ω is defined by $\omega_n = \|f_n\|_2$.
It can be shown that if the sequence (f_n) considered above has the additional property that for each n the function f_n^2 is a multiple of a characteristic function, then $[f_n ; n]$ is complemented in L^p by the ortogonal projection. Because all the $X_{p,\omega}$ can be realized by such sequences, we conclude that the $X_{p,\omega}$ are either $\mathcal{L}^p$-spaces or Hilbert spaces. As shown in [109], there are four cases to distinguish.

PROPOSITION 4.6 :

i) If $\lim \omega_n > 0$, then $X_{p,\omega}$ is isomorphic to l^2.

ii) If $\Sigma_n \omega_n^{\frac{2p}{p-2}} < \infty$, then $X_{p,\infty}$ is isomorphic to l^p.

iii) If we can split the integers $\mathbb{N}$ into two disjoint infinite sets M and N, such that

$$\inf \{\omega_n ; n \in M\} > 0 \quad \text{and} \quad \Sigma_{n\in N} \omega_n^{\frac{2p}{p-2}} < \infty,$$

then $X_{p,\omega}$ is isomorphic to Y_p.

iv) In the other cases, the spaces $X_{p,\omega}$ turn out to be isomorphic to a same space which we denote X_p and which is not complemented in Y_p.

In [95], G. Schechtman obtained infinitely many non-isomorphic separable $\mathcal{L}^p_\lambda$-spaces but with increasing λ.

In this chapter, the existence will be shown of an uncountable family of mutually non-isomorphic complemented subspaces of L^p. So this leads to uncountably many separable $\mathcal{L}^p_\lambda$-types for some $\lambda < \infty$. For some time, one believed that a separable $\mathcal{L}^p$-space either embeds in Z_p or contains $L^p[0,1]$. We will solve this question negatively, by proving that if a separable Banach space B contains a copy of any separable $\mathcal{L}^p$-space without L^p-subspace, then B has a subspace isomorphic to L^p. In particular, the separable $\mathcal{L}^p$-spaces not containing L^p have no universal element.

Besides general Banach space arguments, crucial use is made of classical martingale inequalities due to D. Burkholder, B. Davis, R. Gundi and E. Stein. Also some settheoretical techniques are involved.

2. SOME BASIC PROBABILISTIC FACTS

The structure of L^p depends heavily on certain martingale inequalities. We will explicit here certain basic results which are needed. For proofs and related facts, we refer to [60].

Let (Ω, F, μ) be a fixed probability space and assume $(F_n)_{n=1,2,\ldots}$ an increasing sequence of sub-σ-fields of F. If $f \in L^1(\mu)$, we let $f_n = E[f|F_n]$ and assume $f_0 = 0$. The sequence (f_n) is a martingale and the corresponding martingale difference sequence is given by $\Delta f_n = f_n - f_{n-1}$.

Define for each n

$$f_n^* = \max_{1 \leq k \leq n} |f_k| \quad \text{and} \quad S_n(f) = (\Sigma_{k=1}^n [\Delta f_k]^2)^{1/2}$$

The maximal function f^* and the square function $S(f)$ are given by

$$f^* = \lim_{n\to\infty} f_n^* \quad \text{and} \quad S(f) = \lim_{n\to\infty} S_n(f).$$

We first state Doob's inequality

PROPOSITION 4.7 : For $1 < p < \infty$

$$\|f\|_p \leq \|f^*\|_p \leq \frac{p}{p-1} \|f\|_p$$

The first inequality is of course trivial. The second one follows from a stopping time argument.

Less elementary is the Burkholder-Gundi inequality about the square function.

PROPOSITION 4.8 : Let $1 < p < \infty$. Then there are constants $c_p > 0$ and $C_p < \infty$ such that for all $f \in L^p(\mu)$

$$c_p \|f\|_p \leq \|S(f)\|_p \leq C_p \|f\|_p$$

holds.

We denote (h_i) the L^∞-normalized Haar system on $[0,1]$, i.e. the system of functions $(h_{n,k})_{\substack{n=1,2,\ldots \\ 1\leq k\leq 2^n}}$, where

$$h_{n,k} = \chi_{[(k-1)2^{-n},(2k-1)2^{-n-1}]} - \chi_{[(2k-1)2^{-n-1},k2^{-n}]}$$

and χ means the characteristic function.

It is easily seen that (h_i) is a martingale difference sequence and therefore, by 4.8

PROPOSITION 4.9 : If $1 < p < \infty$, then for any finite sequence (a_i) of scalars

$$c_p \|\Sigma_i \; a_i \; h_i\|_p \leq \|\sqrt{\Sigma_i \; a_i^2 \; h_i^2}\|_p \leq C_p \|\Sigma_i \; a_i \; h_i\|_p$$

holds.

Since $L^p[0,1]$ is generated by the Haarfunctions, we conclude that (h_i) is an unconditional basis for L^p $(1 < p < \infty)$.
Let us remark at this point that the Lebesgue space $[0,1]$ may be replaced by any probability space measure-theoretically isomorphic to $[0,1]$.

If X, Y are Banach spaces then a sequence (x_i) in X and a sequence (y_i) in Y are said to be equivalent provided the map $x_i \leftrightarrow y_i$ induces an isomorphism of $[x_i]$ and $[y_i]$.

Following [76], we call a C-tree over a measure space (Ω, F, μ) a system $(A_{n,k})_{\substack{n=1,2,\dots \\ 1\leq k\leq 2^n}}$ in F satisfying following conditions

(i) $A_{n,k} = A_{n+1,2k-1} \cup A_{n+1,2k}$

(ii) $A_{n+1,2k-1} \cap A_{n+1,2k} = \phi$

(iii) $C^{-1}\, 2^{-n} \leq \mu(A_{n,k}) \leq C\, 2^{-n}$

for each n and $k = 1,\dots,2^n$.

The next result is due to Gamlen and Gaudet (cfr. [58]).

PROPOSITION 4.10 : Let $C < \infty$ and $(A_{n,k})_{n,k}$ a C-tree over the measure space (Ω, F, μ). Consider the sequence $(x_{n,k})_{n,k}$ defined by $x_{n,k} = \chi_{A_{n+1,2k-1}} - \chi_{A_{n+1,2k}}$. Then for all $1 \leq p < \infty$

1. $(x_{n,k})$ in $L^p(\mu)$ is equivalent to $(h_{n,k})$ in $L^p[0,1]$
2. $[x_{n,k}]$ is isomorphic to $L^p[0,1]$
3. $[x_{n,k}]$ is norm-1 complemented in $L^p(\mu)$

We will also need for our purpose the following particular case of a general inequality involving convex functions due to Burkholder, Davis and Gundi.

PROPOSITION 4.11 : Let (Ω, F, μ) be a probability space, (F_n) an increasing sequence of sub-σ-fields of F and (f_n) a sequence of non-negative measurable functions on Ω. Then

$$\| \Sigma_n E[f_n | F_n] \|_p \leq p \| \Sigma_n f_n \|_p$$

holds for all $1 \leq p < \infty$.

We include a proof which is quite simple. It is based on the following elementary calculus lemma left as an exercice to the reader.

LEMMA 4.12 : Let $a_1,\dots,a_N$ be positive numbers and $1 \leq p < \infty$. Then

$$(\Sigma_{n=1}^N a_n)^p \leq p\, \Sigma_{n=1}^N (\Sigma_{j=1}^n a_j)^{p-1} a_n$$

Proof of 4.11 : From the lemma, we get the pointwise inequality

$$(\Sigma_n E[f_n|F_n])^p \leq p\, \Sigma_n(\Sigma_{j=1}^n E[f_j|F_j])^{p-1} E[f_n|F_n].$$

By integration and application of Hölder's inequality, it follows

$$\|\Sigma_n E[f_n|F_n]\|_p^p \leq p\|\Sigma_n(\Sigma_{j=1}^n E[f_j|F_j])^{p-1} f_n\|_1$$

$$\leq p\|(\Sigma_n E[f_n|F_n])^{p-1} (\Sigma_n f_n)\|_1$$

$$\leq p\|\Sigma_n E[f_n|F_n]\|_p^{p-1} \|\Sigma_n f_n\|_p.$$

So $\|\Sigma_n E[f_n|F_n]\|_p \leq p\|\Sigma_n f_n\|_p$, as required.

3. OPERATORS FIXING L^p AND L^p-EMBEDDINGS

Also in this section, we will omit several proofs which are rather long although mostly based on simple ideas. Our main reference here is [76] (Sec. 9, see also remark 1 on p. 263).

Our interest goes first to operators $T : L^p(\mu) \to L^p(\nu)$ which fix an L^p-copy, i.e. which are isomorphisms when restricted to some subspace of $L^p(\mu)$ isomorphic to $L^p[0,1]$.

THEOREM 4.12 :

1. Let $1 \leq p < \infty$ and $p \neq 2$. If $T : L^p(\mu) \to L^p(\nu)$ is an operator fixing a copy of L^p, then there exists a purely non-atomic σ-field $\mathcal{G}$ of μ-measurable sets so that T induces an isomorphism on the subspace $L^p(\mu|\mathcal{G})$ of $L^p(\mu)$ and the image $T(L^p(\mu|\mathcal{G}))$ is a complemented subspace of $L^p(\nu)$.

2. Assume $1 \leq p < 2$. Let $T : L^p(\mu) \to L^p(\nu)$ be an operator and $\delta > 0$ such that $\|Tf\|_p \geq \delta$ whenever f is a μ-measurable function of mean zero taking only the values $+1$ and -1. Then T fixes an L^p-copy.

The case $p = 1$ is the Enflo-Starbird theorem (see [56]). The cases $1 < p < 2$ and $2 < p < \infty$ are proved separately and are due to G. Schechtman (see [76]).

As an immediate consequence of theorem 4.12 (1), we find

COROLLARY 4.13 : If $1 \leq p < \infty$ and X is a subspace of $L^p(\mu)$ isomorphic to L^p, then X has a subspace Y such that Y is also isomorphic to L^p and Y is moreover complemented in $L^p(\mu)$.

Actually, for our purpose, we need the following more precise result, which is in fact the main step in the proof of Th. 4.12. for $1 < p < \infty$ $(p \neq 2)$.

PROPOSITION 4.14 : Let $1 < p < \infty$, $p \neq 2$ and X a subspace of $L^p(\mu)$ which is isomorphic to L^p. Then there exist a system $(h_{n,k})$ in X and a "tree" $(\varphi_{n,k})$ of functions such that the systems $(h_{n,k})$ and $(2^n \varphi_{n,k})$ are biortogonal.

We still have to point out what is meant by a tree of functions. This notion is closely related to trees of sets, which we introduced in previous section.

DEFINITION 4.15 : If (Ω,μ) is a probability space, then a system $\{\varphi_{n,k}$, $n=0,1,2,\ldots$ and $1 \leq k \leq 2^n\}$ of $(0,1,-1)$-valued μ-measurable functions will be called a "tree of functions" provided $(A_{n,k})$ is a C-tree of sets for some $C < \infty$, where $A_{n,k} = \text{supp}\ \varphi_{n,k}$.

Given such a tree $(\varphi_{n,k})$, we define for convenience an "elementary function" as a sum

$$\varphi = \Sigma\ \varepsilon_\alpha\ \varphi_\alpha,$$

where $\varepsilon_\alpha = \pm 1$, the φ_α are $\varphi_{n,k}$-functions with disjoint supports and supp $\varphi = A_{0,1}$.

4. TREES AND TREE-ORDINALS

The aim of this section is to introduae the notion of "tree" and state the Kunen-Martin boundedness result on sets of well-founded trees. In a next section, we will introduce certain subspaces of L^p using trees on the integers. The non-existence of universal elements in this class of spaces is based on ordinal properties of analytic sets of trees. We will suppose the reader familiar with the notion of analytic subset of a Polish space and their stability properties.
Basically we are only interested in trees on the set $\mathbb{N}$ of positive integers. However, also trees of trees will be considered. Therefore, we are obliged to deal with more general trees.

Let us start with an arbitrary set X. The set $\cup_{n=1}^{\infty} X^n$ of the finite complexes of elements of X can be partially ordered in a natural way, by taking $(x_1,\ldots,x_n) < (x_1',\ldots,x_p')$ provided $p \geqslant n$ and $x_k = x_k'$ for $k = 1,\ldots,n$. Comparability and incomparability will always be related to this order. A tree T on X will be a subset of $\cup_{n=1}^{\infty} X^n$ with the property that a predecessor of a member of T belongs also to T. Thus $(x_1,\ldots,x_n) \in T$ whenever $(x_1,\ldots,x_n,x_{n+1}) \in T$.
For a tree T on X, we define
$D(T) = \cup_{n=1}^{\infty} \{(x_1,\ldots,x_n) \in X^n \; ; \; (x_1,\ldots,x_n,x) \in T \text{ for some } x \in X\}$.

Proceeding by induction, we can then construct a transfinite system of trees :
Take $T^0 = T$.
If T^α is obtained, let $T^{\alpha+1} = D(T^\alpha)$.
For limit ordinals γ, define $T^\gamma = \cap_{\alpha<\gamma} T^\alpha$.

The tree T on X is well-founded provided there is no sequence (x_n) in X satisfying $(x_1,\ldots,x_n) \in T$ for each n.
If T is well-founded, then the T^α are strictly decreasing. Hence T^α will be empty for α sufficiently large. The ordinal $o[T]$ of the well-founded tree T will be the smallest ordinal for which $T^{o[T]} = \phi$.

The next result relates the ordinal of a tree to these of certain subtrees.

PROPOSITION 4.16 : Let T be a well-founded tree on X and take $T_x = \bigcup_{n=1}^{\infty} \{(x_1,\dots,x_n) \in X^n \; ; \; (x,x_1,\dots,x_n) \in T\}$ for all $x \in X$. Then

$$o[T] = \sup_{x\in X} (o[T_x] + 1)$$

Proof : It is easily verified by induction α that $(T_x)^\alpha = (T^\alpha)_x$. If $x \in X$ is fixed and $\alpha < o[T_x]$, then $(T_x)^\alpha = (T^\alpha)_x \neq \phi$ and therefore $x \in T^{\alpha+1}$. Distinguishing the cases $o[T_x]$ is not a limit ordinal, $o[T_x]$ is a limit ordinal, we see that $x \in T^{o[T_x]}$ and hence $o[T] \geq o[T_x] + 1$. So $o[T] \geq \sup_{x\in X} (o[T_x] + 1)$.

Let conversely $\alpha = \sup_{x\in X} (o[T_x] + 1)$. For all $x \in X$, we have that $(T_x)^{o[T_x]} = (T^{o[T_x]})_x = \phi$. But this means that no complexes in $T^\alpha \subset T^{o[T_x]+1}$ starts with x. Thus $T^\alpha = \phi$ and $o[T] \leq \alpha$.

A tree T on a Polish (= complete metrizable and separable) space X is said to be analytic provided for all $n \in \mathbb{N}$ the set

$$T_{(n)} = \{(x_1,\dots,x_n) \in X^n \; ; \; (x_1,\dots,x_n) \in T\}$$

is an analytic subset of the product space X^n.

We are now prepared to state the Kunen-Martin theorem which is sometimes also formulated in terms of well-founded relations (cfr. [42]).

THEOREM 4.17 : An analytic tree T on a Polish space X is either not well-founded or $o[T] < \omega_1$.

The result needed for later use is a consequence of 4.1 . We denote $C = \bigcup_{n=1}^{\infty} \mathbb{N}^n$ and $X = \{0,1\}^C$, endowed with the product topology. Thus X is a compact metrizable space. Any tree on $\mathbb{N}$ can be seen as an element of X. The following result holds

COROLLARY 4.18 : If $\mathcal{W}$ is a set of well-founded trees on $\mathbb{N}$ and if $\mathcal{W}$ is an analytic subset of X, then $\sup_{T\in\mathcal{W}} o[T] < \omega_1$.

We will give here only a sketch of the argument. The details can be found in [42] also. For given $\mathcal{W}$, we introduce a new tree $\mathcal{T}$ on X, taking

$$\mathcal{T} = \bigcup_{n=0}^{\infty}\{(T,T_1,\dots,T_n) : T \in \mathcal{W},\ T_1,\dots,T_n \neq \emptyset \text{ are trees on } \mathbb{N},\quad T_i \subset D(T_{i+1}) \text{ for } i=1,\dots,n-1 \text{ and } T_n \subset T\}$$

It is indeed clear that $\mathcal{T}$ is a tree. It is not difficult to verify that if $\mathcal{W}$ consists of well-founded trees then $\mathcal{T}$ is well-founded and moreover using previous notations, $o[\mathcal{T}] > o[\mathcal{T}_T] = o[T]$ for any $T \in \mathcal{W}$. If now $\mathcal{W}$ is analytic in X, it follows from the above definition of $\mathcal{T}$ that $\mathcal{T}$ is an analytic tree on X. Thus $o[\mathcal{T}] < \omega_1$, by 4.17, and this ends the proof.

5. TRANSLATION INVARIANT L^p-EMBEDDINGS

The $\mathcal{L}^p$-spaces obtained in this chapter will appear as translation-invariant subspaces of the Cantor-group. In order to establish non-embeddability of L^p, we will rely on what means analytically an L^p-embedding in this situation. It is the purpose of this section to present a description of sets of characters on the Cantor-group for which the L^p-closed linear span contains an L^p-copy.

Let us first recall some standard terminology.
In what follows, G will denote the Cantor group $\{1,-1\}^{\mathbb{N}}$. The dual group Γ of G is formed by the Walsh functions $w_S = \prod_{n\in S} r_n$, where (r_n) is the Rademacher sequence (= coordinate functions). For $\Lambda \subset \Gamma$ and $1 \leq p < \infty$ we write L^p_Λ for the subspace of $L^p(G)$ generated by the characters $\{\gamma \ ;\ \gamma \in \Lambda\}$.
For $2 < p < \infty$, the sets $\Lambda \subset \Gamma$ for which L^p embeds (in the Banach space sense) in L^p_Λ are characterized by the following theorem

THEOREM 4.19 : Let $\Lambda \subset \Gamma$ and $2 < p < \infty$. Then L^p embeds in L^p_Λ if and only if there exist two sequences (γ_k) and (δ_k) in Γ such that the γ_k are independent Walsh functions and $\gamma.\delta_k$ belongs to Λ for all k and γ in the group generated by $\gamma_1,\gamma_2,\dots,\gamma_k$.

We agree to call (+) the property for subsets $\Lambda \subset \Gamma$ stated in Th. 4.19. The "IF" part is a consequence of the following result.

<u>PROPOSITION 4.20</u> : If $\Lambda \subset \Gamma$ has (+), then there exists $\Lambda' \subset \Lambda$ such that $L^q_{\Lambda'}$ is isomorphic to L^q and complemented in $L^q(G)$, for all $1 < q < \infty$.

<u>Proof</u> : The argument is straightforward. Denote Γ_n (n=0,1,2,...) the diadic (= Littlewood-Paley) decomposition of Γ. Thus

$$\Gamma_0 = \{1\} \text{ and } \Gamma_n = \{w_S \ , \ \max(S) = n\} \text{ for } n = 1,2,\ldots$$

It is clear that (eventually replacing the δ_k) one can assume

$$\Lambda'_k = \{\gamma.\delta_k \ , \ \gamma \text{ is in } [\gamma_1,\ldots,\gamma_k]\} \text{ is contained in } \Gamma_{n_k},$$

where (n_k) is a strictly increasing sequence of integers.
Our aim is to show that

$$\Lambda' = \bigcup_k \Lambda'_k$$

satisfies. Let $\mathcal{E}_k$ be the expectation with respect to the algebra generated by $\{\gamma_1,\ldots,\gamma_k\}$. Notice that if for each k one considers an $\mathcal{E}_k$-measurable function f_k, we get by 4.8 for $1 < q < \infty$

$$\|\Sigma \, f_k.\delta_k\|_q \sim \|(\Sigma \, |f_k|^2)^{1/2}\|_q$$

where $\sim$ means an equivalence depending on q.

Let Θ be the group $[\gamma_k]$ generated by the sequence (γ_k). Since the γ_k are independent we get a trivial isometry of $L^q(G)$ and L^q_Θ.
From the above observation, it follows that the map

$$T : L^q_{\Lambda'} \to L^q_\Theta \quad \text{defined by} \quad T(\gamma.\delta_k) = \gamma.\gamma_{k+1} \text{ for } \gamma \in [\gamma_1,\ldots,\gamma_k]$$

is an isomorphism for $1 < q < \infty$. In particular, $L^q_{\Lambda'}$ and L^a are isomorphic.
It is well-known that the best projection on a translation-invariant subspace is the orthogonal projection. For more details on this matter, the reader may consult [81].

Now the orthogonal projection P from $L^q(G)$ onto L^q_Λ, is given by

$$P(f) = \Sigma_k \ \mathcal{E}_k[f_{n_k} \circ \delta_k] \cdot \delta_k$$

where

$$f = \Sigma f_n \text{ is the diadic decomposition of } f.$$

Applying again 4.8 and 4.11, we find for $q \geq 2$

$$\|P(f)\|_q \sim \|(\Sigma_k \ |\mathcal{E}_k[f_{n_k} \cdot \delta_k]|^2)^{1/2}\|_q$$

$$\lesssim \|(\Sigma_k \ |f_{n_k}|^2)^{1/2}\|_q$$

$$\leq \|(\Sigma \ |f_n|^2)^{1/2}\|_q$$

$$\sim \|f\|_q,$$

establishing the boundedness of P for $2 \leq q < \infty$. The boundedness for $1 < q \leq 2$ follows then by a duality argument. Let us however point out that above estimate is valid for all $1 < q < \infty$, replacing 4.11 by an inequality of E. Stein (cfr. [116]).

The proof of the reverse part of Th. 4.19 is less trivial. In the following crucial definition we consider again trees of functions as introduced in Sec. 3, where the measure space is the Cantor group G equipped with its Haar-measure.

DEFINITION 4.21 : We say that $\Lambda \subset \Gamma$ has property $(*)$ provided for any tree of functions $(\varphi_{n,k})$ on G and $\varepsilon > 0$, there exists a $(\varphi_{n,k})$-elementary function φ for which

$$\left|\int f \varphi \, dm\right| \leq \varepsilon \|f\|_2 \quad \text{for} \quad f \in L^2_\Lambda$$

holds.

The main ingredient of the "only if" part of 4.19 are certain stability properties of $(*)$.

PROPOSITION 4.22 :

(1) The class of subsets Λ of Γ verifying $(*)$ is stable for finite union.

(2) Assume $\Lambda \subset \Gamma$ such that for any finite subset Γ_0 of Γ there exists $\Lambda_0 \subset \Lambda$ such that Λ_0 has (*) and $\gamma.\delta$ is not in Γ_0 whenever γ and δ are distinct elements of $\Lambda \setminus \Lambda_0$.
Then also Λ verifies (*).

Prop. 4.21 suggests the following transfinite system $(S_\alpha)_{\alpha<\omega_1}$ of classes of subset Λ of Γ.

DEFINITION 4.23 :

(1) Denote S_0 the class of 1-element subsets of Γ.

(2) Suppose S_α defined for $\alpha < \beta$. Then S_β is the class of the $\Lambda \subset \Gamma$ such that for all finite $\Gamma_0 \subset \Gamma$, there exists $\Lambda_0 \subset \Lambda$ where Λ_0 is a finite union of elements of S_α for some $\alpha < \beta$ and $\gamma.\delta \notin \Gamma_0$ for $\gamma \neq \delta$ in $\Lambda \setminus \Lambda_0$.

Using 4.22, transfinite induction gives immediately

COROLLARY 4.24 : If $\Lambda \in S_\alpha$ for some $\alpha < \omega_1$, then Λ has (*).

The "regularity" of L^p-embeddings in subspaces of L^p is used to show the following fact.

PROPOSITION 4.25 : If $2 < p < \infty$ and L^p embeds in L^p_Λ, then Λ fails (*).

Proof : Take $(h_{n,k})$ and $(\varphi_{n,k})$ as in prop. 4.14. Consider a $(\varphi_{n,k})$-elementary function

$$\varphi = \Sigma\, \varepsilon_\alpha\, \varphi_\alpha$$

to which we associate the L^p-function

$$f = \Sigma\, \varepsilon_\alpha\, h_\alpha .$$

Notice that since $p > 2$

$$\|f\|_2 \leqslant \|f\|_p \leqslant \text{const.},$$

depending on the equivalence of (h_α) and the Haar-system in L^p.

From the biorthogonality, it follows

$$\int f\,\varphi\,dm = \Sigma_\alpha <h_\alpha,\varphi_\alpha> = 1,$$

because supp $\varphi = A_{0,1}$.
From its behaviour with respect to the tree of functions $(\varphi_{n,k})$, we conclude that Λ fails $(*)$.

For subsets $\Lambda \subset \Gamma$, the properties (+) and $(*)$ are opposite.

PROPOSITION 4.26 : Assume $\Lambda \subset \Gamma$ such that $\Lambda \notin S_\alpha$ for any $\alpha < \omega_1$. Then Λ has property (+).

Proof : We first observe the fact that given a finite subset Γ_f of Γ, there is always a finite partition (Γ_i) of Γ, so that for each i

$$\gamma\delta \notin \Gamma_f \quad \text{whenever} \quad \gamma \neq \delta \text{ in } \Gamma_i$$

(In fact, this property extends to arbitrary groups).
Denote for convenience $S = \bigcup_\alpha S_\alpha$. We show next that if $\Lambda \notin S$, then also $\Lambda \cap \gamma.\Lambda \notin S$ for some $\gamma \neq 1$. Otherwise, since Γ is countable, there should exist $\alpha < \omega_1$ such that

$$\Lambda \cap \gamma.\Lambda \in S_\alpha \quad \text{for all} \quad \gamma \in \Gamma\backslash\{1\}$$

But then $\Lambda \in S_{\alpha+1}$. Indeed, given a finite subset Γ_0 of $\Gamma\backslash\{1\}$, the set

$$\Lambda_0 = \Lambda \cap \bigcup_{\gamma\in\Gamma_0} \gamma\,\Lambda$$

is a finite union of elements of S_α and $\gamma.\delta \notin \Gamma_0$ for γ,δ in $\Lambda\backslash\Lambda_0$.
Proceeding by induction, we will now construct a sequence (γ_k) of distinct elements of Γ and a sequence Λ_k of subsets of Λ, such that

(a) $\Lambda_k \notin S$

(b) $\Lambda_{k+1} \subset \Lambda_k \cap \gamma_{k+1}\Lambda_k$

Start with $\Lambda_0 = \Lambda$. Suppose $\gamma_1,\dots,\gamma_n$ and Λ_n constructed.
Taking $\Gamma_f = \{\gamma_1,\dots,\gamma_n\}$, we get from a preceding observation a finite partition (Γ_i) of Γ, such that for each i and each $k = 1,\dots,n$

$$\gamma.\delta \neq \gamma_k \quad \text{whenever} \quad \gamma \neq \delta \text{ in } \Gamma_i$$

From (a) and since $\Lambda_n = \cup_i (\Gamma_i \cap \Lambda_n)$, there is some i satisfying

$$\Gamma_i \cap \Lambda_n = \Lambda'_n \notin S$$

Again by the preceding

$$\Lambda_{n+1} = \Lambda'_n \cap \gamma_{n+1} \Lambda'_n \notin S \text{ for some } \gamma_{n+1} \in \Gamma \setminus \{1\}.$$

Now $\gamma_{n+1} \notin \Gamma_f$, because for $\gamma \in \Gamma_f \setminus \{1\}$

$$\Lambda'_n \cap \gamma \Lambda'_n \subset \Gamma_i \cap \gamma \Gamma_i = \phi$$

This clearly completes the construction.

Denoting

$$G_k = [\gamma_1, \dots, \gamma_k]$$

an iteration of (b) and (a) show in particular that

$$\bigcap_{\gamma \in G_k} \gamma . \Lambda \neq \phi$$

and thus contains some character δ_k, which will satisfy

$$\gamma \delta_k \in \Lambda \text{ whenever } \gamma \in G_k$$

In order to prove (+), it remains to replace the sequence (γ_k) by a sequence of independent Walshes. Now this procedure is standard (considering suitable products of the γ_k) and left as an exercice to the reader.

We now come back to the proof of Prop. 4.22.

<u>Proof of 4.22 (1)</u>

Denote $\succ$ the diadic order on the set of indices

$$A = \{(n,k) \;,\; n=0,1,2,\dots \text{ and } 1 \leq k \leq 2^n\}$$

Assume Λ, Λ' two (disjoint) subsets of Γ each verifying (*) and fix a tree of functions (φ_α).

For each $(n,k) \in A$, one can consider the "sub-tree"

$$(\varphi_\alpha)_{\alpha \succ (n,k)}$$

Because Λ has (*), there exists a function $\psi_{n,k}$ which is elementary with respect to this sub-tree and such that

$$|\int f\, \psi_{n,k}\, dm| \leq \varepsilon\, 4^{-n-1}\, \|f\|_2 \quad \text{for all} \quad f \in L^2_\Lambda.$$

Now $(\psi_{n,k})$ is again a tree of functions. Since Λ' has $(*)$, there is an $(\psi_{n,k})$-elementary function ψ for which

$$|\int f\, \psi\, dm| \leq \frac{\varepsilon}{2}\, \|f\|_2 \quad \text{if} \quad f \in L^2_{\Lambda'}$$

holds.

Notice that ψ is also (φ_α)-elementary.

If $g \in L^2_{\Lambda\cup\Lambda'}$, orthogonal projection gives

$$g = f + f' \quad \text{where} \quad f \in L^2_\Lambda \quad \text{and} \quad f' \in L^2_{\Lambda'}.$$

We find

$$\begin{aligned} |\int g\, \psi\, dm| &\leq |\int f\, \psi\, dm| + |\int f'\, \psi\, dm| \\ &\leq \Sigma_{n,k}\, |\int f\, \psi_{n,k}|dm + \frac{\varepsilon}{2}\, \|f'\|_2 \\ &\leq \varepsilon(\Sigma_{n,k}\, 4^{-n-1})\|f\|_2 + \frac{\varepsilon}{2}\, \|f'\|_2 \\ &\leq \frac{\varepsilon}{2}\, \|g\|_2 + \frac{\varepsilon}{2}\, \|g\|_2 = \varepsilon\|g\|_2. \end{aligned}$$

Since $\varepsilon > 0$ was chosen arbitrarily, we conclude that $\Lambda \cup \Lambda'$ has $(*)$.

Proof of 4.22 (2)

It is a bit more delicate than (1).

Assume $\Lambda \subset \Gamma$ satisfies the condition stated in 4.22 (2). Let (φ_α) be a tree of functions and (A_α) the corresponding C-tree of supporting subsets of G $(C < \infty)$. Fix also $\varepsilon > 0$.

We fix an integer n, large enough to ensure that $2^{-n} + C\, 2^{-n/2} < \varepsilon$. For $k = 1,\dots,2^n$ denote for convenience χ_k the characteristic function of $A_{n,k}$. Modulo a small perturbation, we can assume

$$\text{Spec}\ \chi_k \subset \Gamma_0 \quad \text{for} \quad k = 1,\dots,2^n,$$

where Γ_0 is some finite subset of Γ.

Now, by hypothesis, there exist $\Lambda_0 \subset \Lambda$ verifying $(*)$ with the property that

$$\gamma.\delta \notin \Gamma_0 \quad \text{for} \quad \gamma \neq \delta \ \text{in}\ \Lambda\setminus\Lambda_0.$$

Next, consider the sub-trees

$$(\varphi_\alpha)_{\alpha \succ (n,1)} \quad (\varphi_\alpha)_{\alpha \succ (n,2)} \quad \cdots \quad (\varphi_\alpha)_{\alpha \succ (n,2^n)}$$

Since Λ_0 has (*), one can find functions ψ_k $(1 \leq k \leq 2^n)$ such that ψ_k is $(\varphi_\alpha)_{\alpha \succ (n,k)}$ elementary and

$$|\int f\, \psi_k\, dm| \leq 4^{-n} \|f\|_2 \quad \text{for} \quad f \in L^2_{\Lambda_0}$$

Moreover, one can construct the ψ_k inductively in such a way that they have essentially disjoint spectrum, thus

$$\text{Spec}\ \psi_k \subset \Phi_k$$

and

$$\Phi_k \cap \Phi_{k'} = \phi \text{ for } k \neq k'.$$

The point is that given a tree of functions on G, one can always recombine the functions in order to obtain a new tree whose members have a spectrum disjoint from a fixed finite subset of Γ. We don't explicit this in full details, since the technique is elementary and standard.

Let P_k the orthogonal projection on Φ_k. Define $\psi = \Sigma\, \psi_k$, which is a (φ_α)-elementary function. We estimate $\int f\, \psi\, dm$ for $f \in L^2_\Lambda$.
If

$$f = f_0 + f_1 \quad \text{with} \quad f_0 \in L^2_{\Lambda_0} \text{ and } f_1 \in L^2_{\Lambda \setminus \Lambda_0}$$

then

$$|\int f\, \psi\, dm| \leq \Sigma_k\, |\int f_0\, \psi_k\, dm| + |\int f_1\, \psi\, dm|$$

$$\leq 2^{-n}\, \|f_1\|_2 + |\int f_1\, \psi\, dm|,$$

from the choice of the ψ_k.
Our purpose is to show that

$$|\int f_1\, \psi\, dm| \leq C\, 2^{-n/2}\, \|f_1\|_2$$

from which it will follow that

$$|\int f\, \psi\, dm| \leq \varepsilon \|f\|_2$$

and thus complete the proof.

We have

$$\int f_1 \psi \, dm = \Sigma_k \int f_1 \psi_k \, dm = \Sigma_k \int P_k(f_1)\psi_k \, dm,$$

hence

$$\left|\int f_1 \psi \, dm\right| \leqslant \Sigma_k \int |P_k(f_1)| \, \chi_k \, dm$$

Remark that if $\gamma \neq \delta$ are in Spec $(P_k(f_1)) \subset (\Lambda \setminus \Lambda_0)$, then

$$\int (\chi_k . \gamma . \delta) \, dm = 0.$$

Consequently

$$\int |P_k(f_1)|^2 \chi_k \, dm = \|P_k(f_1)\|_2^2 \left(\int \chi_k \, dm\right),$$

leading to the estimate

$$\begin{aligned} \left|\int f_1 \psi \, dm\right| &\leqslant \Sigma_k \left(\int \chi_k\right)^{1/2} \left\{\int |P_k(f_1)|^2 \chi_k\right\}^{1/2} \\ &= \Sigma_k \left(\int \chi_k\right) \|P_k(f_1)\|_2 \\ &= \Sigma_k \, m(A_{n,k}) \, \|P_k(f_1)\|_2 \\ &\leqslant C \, \Sigma_k \, 2^{-n} \, \|P_k(f_1)\|_2 \\ &\leqslant C \, 2^{-n/2} \, \|f_1\|_2, \end{aligned}$$

as required.

So prop. 4.22 is established.

Combining 4.25, 4.24, 4.26, the "ONLY IF" part of 4.19 is obtained.

Although we don't know if 4.19 goes true for $1 < p < 2$, dualization leads to

COROLLARY 4.27 : For $\Lambda \subset \Gamma$, following properties are equivalent

(1) Λ has (+)

(2) There exist $\Lambda_0 \subset \Lambda$ and $1 < p < \infty$ $(p \neq 2)$ such that $L^p_{\Lambda_0}$ is complemented in $L^p(G)$ and contains a copy of L^p.

Proof : That (1) ⇒ (2) is a consequence of 4.20. Conversely, we can conclude by 4.19 in case $2 < p < \infty$. Now if $1 < p < 2$, we first use 4.13 to obtain a complemented embedding of L^p in $L^p_{\Lambda_0}$. By duality, $L^{p'}$ embeds in $(L^p_{\Lambda_0})^*$. Next, from the hypothesis and the fact that orthogonal projection is self-dual, it follows that $(L^p_{\Lambda_0})^*$ is isomorphic to $L^{p'}_{\Lambda_0}$. Because $2 < p' < \infty$, application of 4.19 ends the proof.

Let us also remark following "primarity" property of (+)

COROLLARY 4.28 : If $\Lambda \subset \Gamma$ has (+) and $\Lambda = \Lambda' \cup \Lambda''$, then either Λ' or Λ'' will have (+).

Proof : Λ satisfies the conclusion of 4.20 and hence, by 4.25, Λ fails (*).
Next, by 4.24, $\Lambda \notin S_\alpha$ for all $\alpha < \omega_1$. Thus this will also be true for either Λ' or Λ''. It remains to use 4.26.

4.28 is a purely combinatorial result. A direct proof will be indicated in the remarks at the end of this chapter.

Several results of this section extend to other groups than the Cantor group. More details on this matter will be given later.

6. A COMPLEMENTED SUBSPACE OF L^p

Again we let $C = \bigcup_{n=1}^\infty \mathbb{N}^n$. Consider the group $G = \{-1,1\}^C$ equipped with the Haar measure, which may obviously be identified with the Cantor group. For all $c \in C$, the Rademacker function r_c on G is given by $r_c(x) = x(c)$. To each finite subset F of C corresponds a Walsh function $w_F = \Pi_{c\in F}\, r_c$. This system of Walsh functions generates $L^p(G)$ for all $1 \leq p < \infty$.

We say that a measurable function f on G only depends on the coordinates $F \subset C$ provided $f(x) = f(y)$ whenever $x,y \in G$ with $x(c) = y(c)$ for all $c \in F$. A measurable subset S of G depends only on the coordinates $F \subset C$ provided χ_S does.
For $F \subset C$, denote $\mathcal{G}(F)$ the sub-σ-algebra of those subsets of G only depending on the F-coordinates.
Remark that if $F' \subset C$, $F'' \subset C$ then $\mathcal{G}(F') \cap \mathcal{G}(F'') = \mathcal{G}(F' \cap F'')$. The conditional expectations with respect to $\mathcal{G}(F')$ and $\mathcal{G}(F'')$ commute and their composition gives the expectation with respect to $\mathcal{G}(F' \cap F'')$.

A branch in C will be a subset of C consisting of mutually comparable elements. The following definition is crucial.

Let $1 \leq p < \infty$. We denote X_C^p the closed linear span in $L^p(G)$ over all finite branches Γ in C of all those functions in $L^p(G)$ which depend only on the coordinates of Γ.
Thus X_C^p is the subspace of $L^p(G)$ generated by the set of Walsh-functions $\{w_\Gamma \,;\, \Gamma \text{ is a finite branch in } C\}$.

<u>THEOREM 4.30</u> : X_C^p is complemented in $L^p(G)$ for all $1 < p < \infty$.

The proof of this result is the main objective of this section.

Let us say that a set $(\mathcal{E}_i)$ of sub-σ-algebras of G is compatible if for all i and j either $\mathcal{E}_i \subset \mathcal{E}_j$ or $\mathcal{E}_j \subset \mathcal{E}_i$ holds. It is evident that prop. 4.11 holds for compatible sequences $(\mathcal{E}_i)$ as well. Indeed, fix non-negative measurable functions $f_1,\dots,f_n$. The compatibility of the $\mathcal{E}_i$'s implies that there is a permutation σ of $\{1,\dots,n\}$ with $\mathcal{E}_{\sigma(i)} \subset \mathcal{E}_{\sigma(j)}$ for all $1 \leq i < j \leq n$. Hence

$$\|\Sigma_i\, E[f_i|\mathcal{E}_i]\|_p = \|\Sigma_i\, E[f_{\sigma(i)}|\mathcal{E}_{\sigma(i)}]\|_p$$

$$\leq p\, \|\Sigma_i\, f_{\sigma(i)}\|_p = p\, \|\Sigma_i\, f_i\|_p .$$

In fact, it will be shown that X_C^p is complemented in $L^p(G)$ by the ortogonal projection P and at the same time p-norm estimates on P will be obtained.

It is sufficient to treat the case $2 \leq p < \infty$, since clearly the conjugate P^* of P is the ortogonal projection on $X_C^{p'}$ in $L^{p'}(G)$ $(p' = \frac{p}{p-1})$.

By the nature of the property we have to deal with, $\mathcal{C}$ may be as well replaced by a finite subtree $\mathcal{C}_0$ of $\mathcal{C}$, for instance

$$\mathcal{C}_0 = \cup_{n=1}^{N} \{1,2,\ldots,N\}^n$$

for some positive integer N.
We need an explicit order-preserving enumaration γ of $\mathcal{C}_0$.
For $c = (c_1,\ldots,c_n) \in \mathcal{C}_0$, put

$$\gamma(c) = \frac{N^n - N}{N - 1} + \Sigma_{i=1}^{n} (c_i - 1) N^{n-i} + 1$$

Thus the enumeration is as follows

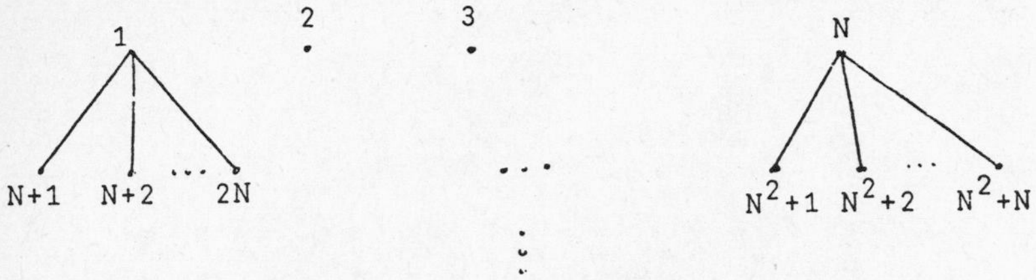

Next, we introduce 3 systems of sub-σ-algebra's.

For $i = 0,1,\ldots,\frac{N^{N+1} - N}{N - 1}$, take $F_i = \mathfrak{S}(c \in \mathcal{C}_0 \ ; \ \gamma(c) \leq i)$.

For all maximal complexes c in $\mathcal{C}_0$ ($c \in \mathcal{C}_0$, $|c| = N$), define

$\mathcal{E}'_c = \mathfrak{S}(F'_c)$

where

$F'_c = \cup_{n=1}^{N} \{d \in \mathcal{C}_0 \ ; \ |d| = n \text{ and } \gamma(d) \leq \gamma(c|n)\}$

$\mathcal{E}''_c = \mathfrak{S}(F''_c)$

where

$F''_c = \cup_{n=1}^{N} \{d \in \mathcal{C}_0 \ ; \ |d| = n \text{ and } \gamma(d) \geq \gamma(c|n)\}$

It is easily verified that $(F_i)_i$ is increasing and both families $(\mathcal{E}'_c)_{c \text{ maximal}}$ and $(\mathcal{E}''_c)_{c \text{ maximal}}$ are compatible.

We will now express the orthogonal projection P_0 on $X^p_{C_0}$.

To do this, we need "branch-algebras".
Take B_ϕ = trivial algebra and $B_c = \mathcal{G}(d \in C_0 \ ; d \leq c)$ for each $c \in C_0$.
For $c \in C_0$ and $|c| = 1$, let $c' = \phi$. For $c \in C_0$ and $|c| > 1$, let c' be the predecessor of c in C_0.
It is not difficult to see that P_0 is given by

$$P_0(f) = E[f|B_\phi] + \Sigma_{c\in C_0}(E[f|B_c] - E[f|B_{c'}]) \qquad \text{(i)}$$

Take $c \in C_0$ and let $i = \gamma(c)$. Clearly $E[f|B_c] - E[f|B_{c'}]$ is F_i-measurable and $E[\,E[f|B_c] - E[f|B_{c'}] \mid F_{i-1}] = 0$.
Indeed, by the observations made at the beginning of this section,
$B_c \cap F_{i-1} = B_{c'} = B_{c'} \cap F_{i-1}$
and
$E[\,E[f|B_c] \mid F_{i-1}] = E[f|B_{c'}] = E[\,E[f|B_{c'}] \mid F_{i-1}]$

So in (i), $P_0(f)$ is written as a sum of a martingale difference sequence. Application of the Burkholder-Gundi inequality (see 4.8) yields the estimate

$$\|P_0 f\|_p \leq c_p^{-1} \|\{E[f|B_\phi]^2 + \Sigma_{c\in C_0}(E[f|B_c] - E[f|B_{c'}])^2\}^{1/2}\|_p \qquad \text{(ii)}$$

So it remains to estimate the right side of (ii).

By the reverse Burkholder-Gundi inequality, we get

$$\|g_0^2 + \Sigma_{i\geq 1} g_i^2\|_{\frac{p}{2}} \leq C_p^2 \|f\|_p^2 \qquad \text{(iii)}$$

taking $g_0 = E[f|B_\phi]$ and $g_i = E[f|F_i] - E[f|F_{i-1}]$ for $i \geq 1$.

To each $i = 1,\ldots,\frac{N^{N+1} - N}{N - 1}$, we associate a maximal complex $\iota(i)$ in C_0 which succeeds to $\gamma^{-1}(i)$, thus such that $\gamma^{-1}(i) \leq \iota(i)$.
By a previous remark, both sequences $(\mathcal{E}'_{\iota(i)})_i$ and $(\mathcal{E}''_{\iota(i)})_i$ are compatible.

Now we always have that $E[g|\mathcal{E}]^2 \leq E[g^2|\mathcal{E}]$ and for each i
$E[E[g_i|\mathcal{E}'_{\iota(i)}] \mid \mathcal{E}''_{\iota(i)}] = E[g_i|\mathcal{E}'_{\iota(i)} \cap \mathcal{E}''_{\iota(i)}]$.

Since $p \geq 2$ it follows from two successive applications of the Burkholder-Davis-Gundi inequality (4.11) that

$$\|g_0^2 + \Sigma_{i\geq 1} E[g_i|\mathcal{E}'_{\iota(i)} \cap \mathcal{E}''_{\iota(i)}]^2\|_{\frac{p}{2}} \leq (\tfrac{p}{2})^2 \|g_0^2 + \Sigma_{i\geq 1} g_i^2\|_{\frac{p}{2}} \qquad \text{(iv)}$$

Combining (iii) and (iv), we find

$$\|g_0^2 + \Sigma_{i\geq 1} E[g_i|\mathcal{E}'_{\iota(i)} \cap \mathcal{E}''_{\iota(i)}]^2\|_{\frac{p}{2}} \leq (\tfrac{p}{2})^2\, C_p^2\, \|f\|_p^2 \qquad \text{(v)}$$

Now for each i holds

$$\mathcal{E}'_{\iota(i)} \cap \mathcal{E}''_{\iota(i)} = \mathcal{G}(F'_{\iota(i)} \cap F''_{\iota(i)}) = \mathcal{G}(d \in C_0\ ;\ d \leq \iota(i)) = \mathcal{B}_{\iota(i)}$$

and taking $c \in C_0$ with $\gamma(c) = i$

$$\mathcal{B}_{\iota(i)} \cap F_i = \mathcal{G}(d \in C_0\ ;\ d \leq \iota(i) \text{ and } \gamma(d) \leq i) = \mathcal{G}(d\in C_0\ ;\ d\leq c) = \mathcal{B}_c$$

$$\mathcal{B}_{\iota(i)} \cap F_{i-1} = \mathcal{G}(d \in C_0\ ;\ d\leq\iota(i) \text{ and } \gamma(d) < i) = \mathcal{G}(d\in C_0\ ;\ d\leq c') = \mathcal{B}_{c'}$$

Hence $E[g_i|\mathcal{E}'_{\iota(i)} \cap \mathcal{E}''_{\iota(i)}] = E[g_i|\mathcal{B}_{\iota(i)}] =$

$$E[f|F_i \cap \mathcal{B}_{\iota(i)}] - E[f|F_{i-1} \cap \mathcal{B}_{\iota(i)}] = E[f|\mathcal{B}_c] - E[f|\mathcal{B}_{c'}]$$

So (v) leads to

$$\|E[f|\mathcal{B}_\phi]^2 + \Sigma_{c\in C_0}(E[f|\mathcal{B}_c] - E[f|\mathcal{B}_{c'}])^2\|_{\frac{p}{2}} \leq (\tfrac{p}{2})^2\, C_p^2\, \|f\|_p^2$$

Consequently, by (ii), we get

$$\|P_0(f)\|_p \leq \frac{p}{2}\,\frac{C_p}{c_p}\,\|f\|_p$$

Hence, the following result holds, which proves in particular 4.19

<u>PROPOSITION 4.31</u> : For all $1 < p < \infty$, the ortogonal projection P on X_C^p is bounded in L^p-norm and more precisely

$$\|P\|_p \leq \frac{p}{2(p-1)}\,\frac{C_{p'}}{c_{p'}} \quad (p' = \frac{p}{p-1}) \quad \text{if } 1 < p \leq 2$$

and

$$\|P\|_p \leq \frac{p}{2}\,\frac{C_p}{c_p} \qquad\qquad \text{if} \quad 2 \leq p < \infty$$

Obviously, to any infinite branch in C corresponds a complemented (by conditional expectation L^p-subspace of $L^p(G)$. So also the subspace X^p_C contains complemented L^p-copies. By this observation and the result 4.30, it follows from Pelczynski's decomposition method (see [87])

COROLLARY 4.32 : For $1 < p < \infty$, the space X^p_C is isomorphic to L^p.

We conclude this section with some remarks.

REMARKS

1. In the proof of (4.30), we may replace again the Burkholder-Davis-Gundi inequality by the following result due to E. Stein (see [116] for the proof).

PROPOSITION 4.33 : Assume $\mathcal{E}_1 \subset \mathcal{E}_2 \subset \ldots$ increasing σ-algebras and $f_1, f_2, \ldots$ arbitrary measurable functions. Then

$$\|\{\Sigma_i (E[f_i|\mathcal{E}_i])^2\}^{1/2}\|_p \leq A_p \|\{\Sigma_i\ f_i^2\}^{1/2}\|_p$$

holds for all $1 < p < \infty$, where A_p depends only on p.

If we use the result, we don't have to use a duality argument in order to show the p-boundedness of the orthogonal projection on X^p_C if $1 < p < 2$. The estimate on the norm is similar, since in fact A_p has order of magnitude $p^{1/2}$ as $p \to \infty$.

2. A natural question which arises when we look at the proof of 4.20 is the following problem

Let $\mathcal{E}_1, \mathcal{E}_2, \ldots$ be a given sequence of sub-σ-algebras of the measure space. Under what (combinatorial) conditions on the $\mathcal{E}_i$'s, is it true that there exists a constant $c_p < \infty$ so that

$$\| \Sigma_i \, E[f_i | \mathcal{E}_i] \|_p \leq c_p \| \Sigma_i \, f_i \|_p$$

for all non-negative measurable functions $f_1, f_2, \ldots$ and $1 \leq p < \infty$?

3. Emphasizing the group structure of $G = \{-1,1\}^C$, it is clear from the definition that X_C^p is a translation invariant subspace of $L^p(G)$.

7. TREE-SUBSPACES OF L^p

L^p is identified to the space $L^p(G)$ with $G = \{-1,1\}^C$. In the previous section, we introduced the space X_C^p, which is complemented in $L^p(G)$ by the ortogonal projection and isomorphic to L^p for $1 < p < \infty$.

A tree T on $\mathbb{N}$ is seen as a subset of C. For $1 \leq p < \infty$, we define the subspace X_T^p of $L^p(G)$ as follows

X_T^p is the closed linear span in $L^p(G)$ over all finite branches Γ in T of all those functions in $L^p(G)$ which depend only on the coordinates of Γ.

So X_T^p is a subspace of X_C^p and is obviously complemented in X_C^p by the expectation with respect to the σ-algebra $\mathfrak{G}(T)$. Combining this fact and 4.19, it follows

THEOREM 4.34 : If T is a tree on $\mathbb{N}$, then X_T^p is a complemented subspace of $L^p(G)$ for all $1 < p < \infty$.

Let us next study the Banach space properties of the X_T^p.

THEOREM 4.35 : For $1 < p < \infty$ and $p \neq 2$, the following holds

1. X_T^p is a $\mathcal{L}^p$-space whenever $o[T]$ is infinite.
2. L^p does not embed in X_T^p if and only if T is well-founded.

Proof :

1. From 4.34 and 4.1, we deduce that X_T^p is either a Hilbertspace or a $\mathcal{L}^p$-space. It is however easily seen that X_T^p contains $\ell^p(n)$-spaces and hence an ℓ^p-copy provided o[T] is infinite. By 1.25, this proves (1).

2. If T contains an infinite branch, then obviously L^p embeds in X_T^p. We have to show that if T is well-founded, then L^p does not imbed in X_T^p. Now X_T^p appears as complemented, translation-invariant subspace of $L^p(G)$ and we can thus apply the results of Section 5. More precisely $X_T^p = L_\Lambda^p$ where $\Lambda = \{w_F$; F is a branch-set contained in T$\}$. By 4.27, it remains to show that Λ fails (+) or, equivalently, Λ belongs to some class S_α $(\alpha < \omega_1)$. It is a simple exercice to verify inductively that in fact $\Lambda \in S_{o[T]}$.

This completes the proof of Th. 4.35.

In [30] the following result about complemented embeddings of L^p into spaces with unconditional Schauder decomposition is established.

THEOREM 4.36 : Let $1 < p < \infty$ and suppose L^p isomorphic to a complemented subspace of a Banach space X with an unconditional Schauder decomposition (X_i). Then one of the following holds

1. There is some i so that L^p is isomorphic to a complemented subspace of X_i
2. A block basic sequence of the X_i's is equivalent to the Haar-system of L^p and has closed linear span complemented in X.

A sequence (b_j) in X is called a block basic sequence of the X_i's if there exist elements $x_i \in X_i$ and integers $n_1 < n_2 < \dots$ with $b_j = \Sigma_{i=n_j+1}^{n_{j+1}} x_i$.

At the end of this chapter we present a proof of Th. 4.36 in case $1 < p < 2$.

From 4.36, we get an alternative approach to 4.35 (2), more in the spirit of general Banach space theory.

Alternative proof of 4.35 (2)

By 4.13, 4.34 and duality we can restrict ourselves to the case $2 < p < \infty$. We proceed by induction on $o[T]$.

Using the notations of section 4, we have that

$$T = \cup_n (n, T_n), \text{ where } (n,T_n) = \{(n,c) \; ; \; c \in T_n\}$$

and

$$o[T] = \sup_n (o[T_n] + 1)$$

The space X_T^p is generated by the sequence of probabilistically mutually independent spaces $B_n = X^p_{(n,T_n)}$. In particular, $\oplus_n B_n$ is an unconditional decomposition of X_T^p. Assume L^p embeds complementably in X_T^p. Application of 4.36 then leads to the following alternative

A. There is some n such that L^p is isomorphic to a complemented subspace of B_n.

B. There is a block basic sequence (b_r) of the B_n's which is equivalent to the Haar system of L^p.

Assume (A) : It is easily seen that B_n is isomorphic to $X^p_{T_n} \oplus X^p_{T_n}$. So by another application of 4.36, L^p should embed complementably in $X^p_{T_n}$. This however is impossible by induction hypothesis and since $o[T_n] < o[T]$.

Assume (B) : A block basic sequence of the B_n's is a sequence of probabilistically independent functions. Since we assumed $2 \leq p < \infty$, it follows from 4.5 that the span of this sequence is isomorphic to a space $X_{p,\omega}$ and these don't contain an L^p-copy (cfr. 4.6).

Our next aim is to show that L^p is the only universal space for the family $(X_T^p)_T$ well-founded.

We recall that a Banach space B is said to be universal for a class K of Banach spaces provided any member of K embeds (isomorphically) in B.

THEOREM 4.37 : If $1 \leq p < \infty$ and B is a separable Banach space which is universal for the class $\{X_T^p$; T well-founded tree on $\mathbb{N}\}$, then B contains a copy of L^p.

Proof : It is of settheoretical nature. Assume B with the above property. Define for each $\delta > 0$

$$\mathcal{W}_\delta = \{T \subset \mathcal{C} \text{ ; T is a tree and there exists a linear operator } \phi : X_T^p \to B \text{ satisfying } \|\phi\| \leq 1 \text{ and } \|\phi(f)\| \geq \delta\|f\| \text{ if } f \in X_T^p\}$$

By a standard argument, we see that for some $\delta > 0$ the set $\mathcal{W}_\delta$ will contain well-founded trees of arbitrarily large ordinal.
Our purpose is to show that $\mathcal{W}_\delta$ is an analytic subset of $\{0,1\}^{\mathcal{C}}$.

Because X_T^p is norm-1 complemented in $X_{\mathcal{C}}^p$, we can also write

$$\mathcal{W}_\delta = \{T \subset \mathcal{C} \text{ ; T is a tree and there exists a linear operator } \psi : X_{\mathcal{C}}^p \to B \text{ satisfying } \|\psi\| \leq 1 \text{ and } \|\psi(f)\| \geq \delta\|f\| \text{ if } f \in X_T^p\}.$$

The set $\mathbf{T}$ of all trees on $\mathbb{N}$ is a closed subspace of $\{0,1\}^{\mathcal{C}}$.
Define

$$\Psi = \{\psi \in \mathcal{L}(X_{\mathcal{C}}^p, B) \text{ ; } \|\psi\| \leq 1\}$$

and endow Ψ with the pointwise topology.
Since B is a separable Banach space, it is clear that Ψ is Polish (for details, see [14]).
We introduce the following subset $\mathcal{S}$ of $\mathbf{T} \times \Psi$

$$\mathcal{S} = \{(T,\psi) \in \mathbf{T} \times \Psi \text{ ; } \|\psi(f)\| \geq \delta\|f\| \text{ if } f \in X_T^p\}$$

We claim that S is Borel with respect to the product-topology. To see this, fix a dense sequence (f_k) in X^p_C. Clearly

$\|\psi(f)\| \geq \delta\|f\|$ for $f \in X^p_T$

and

$\|\psi(f_k)\| \geq \delta\|f_k\| - 2 \underset{\|\ \|}{\text{dist}} (f_k, X^p_T)$ for $k = 1,2,\ldots$

are equivalent conditions for any $T \in \mathbf{T}$ and $\psi \in \Psi$.

Moreover, the map $\mathbf{T} \to \mathbb{R} : T \mapsto \underset{\|\ \|}{\text{dist}} (f, X^p_T)$ is Borel-measurable for any $f \in X^p$. This leads to the required conclusion.

Since now W_δ is the projection of S on $\mathbb{T}$, it follows that W_δ is analytic.
Thus 4.18 applies. By hypothesis $\sup_{T \in W_\delta} o[T] = \omega_1$ and therefore W_δ contains a non-well-founded tree T_0. Because L^p embeds in $X^p_{T_0}$ (cfr. 4.35), we conclude that B contains a copy of L^p.

From 4.35 and 4.37 we deduce the following two consequences

<u>COROLLARY 4.38</u> : The class of complemented subspaces of L^p not containing a copy of L^p has no universal element $(1 < p < \infty, p \neq 2)$

<u>COROLLARY 4.39</u> : There exists an uncountable family of mutually non-isomorphic complemented subspaces of L^p $(1 < p < \infty, p \neq 2)$.

8. RELATED REMARKS AND PROBLEMS

1. Another (combinatorial proof of 4.28 follows from a result of K.R. Milliken (cfr. [92]).

PROPOSITION 4.40 : Given an abelian group G and a subset X of G, denote $<X>^1$ the simple products of elements of X. If now V is an infinite sequence of distinct elements of G and

$$<V>^1 = A_0 \cup A_1 \cup \ldots \cup A_{r-1},$$

then there exists a sequence X in G such that $<X>^1 \subset A_k$ for some $k = 0,1,\ldots,r-1$.

In fact [92] extends work of N. Hindman (see [67]).

2. The "ONLY IF" part of 4.19 easily generalizes to compact abelian groups. However (+) do not imply an L^p-embedding in general. In fact, there are examples of subsets X of $\mathbb{Z}$ for which L^p does not embed in L^p_Λ, where $\Lambda = <X>^1$.

3. The preceding section solves the following question of A. Pelczynski affirmatively for the Cantor group :
Let Γ be an infinite compact abelian group and $1 < p < \infty$, $p \neq 2$. Are there uncountably many non-isomorphic complemented translation-invariant subspaces of $L^p(\Gamma)$. What for the circle group ?

4. For $1 < p < \infty$, $p \neq 2$, the system $(X^p_T)_{T \text{ well-founded}}$ is in some sense the L^p-analogue of the spaces of continuous functions on a countable compact topological space. It should be interesting to clarify the question if there are only $\aleph_1$ isomorphism-types in the system mentioned above.

5. Let us consider the system $(X^1_T)_{T \text{ well-founded}}$ of subspaces of L^1.

Each of these spaces possesses the Radon-Nikodým property. To show this, we proceed of course by induction on o[T]. We use the fact that any unconditional decomposition in L^1 is boundedly complete (L^1 has cotype) and that if $X = \oplus_i X_i$ is a boundedly complete decomposition where each of the spaces X_i has the RNP, then also X has the RNP. Combination of this and 4.37 leads to the following result.

PROPOSITION 4.41 : The class of the subspaces of L^1 possessing the RNP has no universal element.

6. Fix $1 \leq p < \infty$. Proceeding by induction we introduce as follows a system $(R^p_\alpha)_{\alpha<\omega_1}$ of subspaces of L^p :

Let R^p_0 be the one dimensional space of the constant functions.
If R^p_α has been defined, we let $R^p_{\alpha+1}$ equal the L^p-direct sum in L^p of R^p_α with itself.
If α is a limit ordinal and R^p_β has been defined for all $\beta < \alpha$, we let R^p_α equal the independent L^p-sum in L^p of the R^p_β's for $\beta < \alpha$.

It is not difficult to show that for each $\alpha < \omega_1$ one can find a well-founded tree T_α on $\mathbb{N}$ such that $o[T_\alpha] = \alpha$ and R^p_α and $X^p_{T_\alpha}$ are the same spaces. We leave this as an interesting exercice for the reader.
So in particular the spaces $(R^p_\alpha)_\alpha$ are $\mathcal{L}^p$-spaces for $1 < p < \infty$ and L^p is the only universal space for the system $(R^p_\alpha)_\alpha$.
We don't see a direct way (by induction on α for instance) to prove that the R^p_α are complemented in L^p. The tree-representation of these spaces seems to be crucial there.

One may see in R^p_α the L^p-analogue of the space $C(\alpha)$ of continuous functions on the ordinal α. From where the following question
If $1 < p < \infty$ and $p \neq 2$, under what condition on $\alpha < \omega_1$ and $\beta < \omega_1$ are the spaces R^p_α and R^p_β isomorphic ?

Let us remark that this condition will be certainly different than that for the $C(\alpha)$-spaces. For instance, R^p_ω and $R^p_{\omega+\omega}$ are not isomorphic. Indeed, for $2 < p < \infty$, one can verify that R^p_ω is in fact isomorphic to the space X_p and thus a subspace of Y_p. On the other hand it is easily seen that Z_p embeds in $R^p_{\omega+\omega}$.

7. Using again induction on $o[T]$, one can show that the spaces X_T^p are spanned by a martingale difference sequence and hence have an unconditional basis for $1 < p < \infty$.
It is an open question whether or not any $\mathcal{L}^p$-space ($1 < p < \infty$, $p \neq 2$) has an unconditional basis.

8. Is it true that for $1 < p < \infty$ and $p \neq 2$ any $\mathcal{L}^p$-space not containing L^p embeds in some space R_α^p ($\alpha < \omega_1$) ?

9. Are there for all $1 < p < \infty$, $p \neq 2$ a continuum number of separable $\mathcal{L}^p$-types ? This question is related to remark 4

10. The "tree-technique" turns out to be an extremaly powerful way of proving the non-existence of certain universal spaces. For related results on this matter, the reader is referred to [14] and [15].

11. A detailed proof of 4.36 can be found in [30]. This result extends the primarity property of L^p ($1 \leqslant p \leqslant \infty$). Thus if $L^p \sim X \oplus Y$, then either X or Y is an isomorph of L^p. In case $1 < p < 2$, Th. 4.12 (2) provides a simpler argument than presented in [30].

We start with the following lemma

<u>LEMMA 4.42</u> : If $S : L^p(\mu) \to L^p(\nu)$ and $T : L^p(\mu) \to L^p(\nu)$ are operators such that the operator $S + T$ is an into-isomorphism, then one of the operators S, T fixes a copy of L^p.

<u>Proof</u> : Let $\rho > 0$ satisfy $\|(S+T)(f)\|_p \geqslant \rho\|f\|_p$ for any $f \in L^p(\mu)$. Assume that S does not fix L^p. By application of 4.12 (2), we can then conclude the following :
For any μ-measurable set A of positive measure and $\delta > 0$, there exists a measurable function h of mean 0 which is supported by A, only takes the values ± 1 on A and such that $\|S(h)\| < \delta$.
Proceeding by induction we can then construct a 1-tree $(A_{n,k})_{\substack{n=1,2,\dots \\ 1\leqslant k\leqslant 2^n}}$ of μ-measurable sets with the property that
$\|S(h_{n,k})\|_p < \frac{\rho}{2}\, 8^{-n}$ for each $n \in \mathbb{N}$ and $1 \leqslant k \leqslant 2^n$, where

$$h_{n,k} = \chi_{A_{n+1,2k-1}} - \chi_{A_{n+1,2k}}.$$

By prop. 4.10, the system $(h_{n,k})$ is equivalent to the usual Haar system of L^p. We claim that T is an isomorphism when restricted to $[h_{n,k}]$. Indeed, by the hypothesis and the construction, we find for any $f = \Sigma_{n,k}\, a_{n,k}\, h_{n,k}$

$$\begin{aligned}\rho\, \|f\|_p &\leq \|S(f)\|_p + \|T(f)\|_p \leq \\ &\Sigma_{n,k}\, |a_{n,k}|\, \|S(h_{n,k})\|_p + \|T(f)\|_p \leq \\ &\frac{\rho}{2} \max \{2^{-n}\, |a_{n,k}| \ ;\ n=1,2,\ldots\ ;\ 1\leq k\leq 2^n\} + \|T(f)\|_p \leq \\ &\frac{\rho}{2}\, \|f\|_p + \|T(f)\|_p\end{aligned}$$

and thus

$$\|T(f)\|_p \geq \frac{\rho}{2}\, \|f\|_p .$$

A consequence of the preceding lemma is the following fact

LEMMA 4.43 : Assume $T_i : L^p(\mu) \to L^p(\nu)$ an operator for each $i = 1,\ldots,j$. If non of the T_i fixes a copy of L^p, then the operator $\Sigma_{i=1}^{j}\, T_i$ will not fix a copy of L^p.

Let us now come back to theorem 4.36. Let $P_i : X \to X_i$ be the projection according to the decomposition $X = \oplus_i X_i$. Assume Y a complemented subspace of X which is isomorphic to L^p and denote $U : X \to Y$ the projection and $I : L^p \to Y$ the isomorphism.
For each i, we may introduce the operator $T_i : L^p \to L^p$ given by $T_i = I^{-1}\, U\, P_i\, I$

$$L^p \xrightarrow{\ I\ } Y \hookrightarrow X \xrightarrow{\ P_i\ } X_i \hookrightarrow X \xrightarrow{\ U\ } Y \xrightarrow{\ I^{-1}\ } L^p$$

Remark that $\Sigma_i\, T_i$ is the identity operator on L^p.

Suppose now that some operator T_i fixes an L^p-copy. Then, by 4.12 (1), there is a subspace Z of L^p such that Z is isometric to L^p, the restriction $T = T_i|Z$ is an into-isomorphism and $T_i(Z)$ is complemented in L^p by a projection V. Consequently $P_i\ I\ (Z)$ is a subspace of X_i isomorphic to L^p and complemented by the operator $P_i\ I\ T^{-1}\ V\ I^{-1}\ U$.

Thus if non of the X_i has a complemented L^p-subspace, then non of the operators T_i will fix an L^p-copy. Hence, by 4.42, also non of the operators $\Sigma_{i=1}^{j} T_i$ (j=1,2,...). Again from 4.12 (2), the following is true .

If A has positive measure in [0,1], then for any integer j and $\delta > 0$, there exists a Lebesgue-measurable function h of mean 0 which is supported by A, only takes the values ±1 on A and such that $\|\Sigma_{i=1}^{j} T_i(h)\| < \delta$.

In the same spirit as lemma 4.42, a routine construction allows us to obtain a sequence (h_l) in L^p and an increasing sequence (n_l) of integers, so that following properties are satisfied

(i) (h_l) is equivalent to the usual Haar system of L^p and $[h_l]$ is norm-1 complemented in L^p by a projection V

(ii) $\|\Sigma_{i>n_l} P_i\ I\ (h_l)\| < \varepsilon_l\ \|h_l\|_p$

(iii) $\|\Sigma_{i\leq n_l} T_i\ (h_{l+1})\| < \varepsilon_{l+1}\ \|h_{l+1}\|_p$

where $\varepsilon_l^{-1} = 4^l\ (\|I\| + \|U\|)\ \|I^{-1}\|$.

Take $n_0 = 0$ and define $b_l = \Sigma_{i=n_{l-1}+1}^{n_l} P_i\ I\ (h_l)$ for each integer l.

We claim that (h_l) and (b_l) are equivalent sequences and $[b_l]$ is complemented in X. This will conclude the proof of 4.36.

Because $h_l = \Sigma_i T_i(h_l)$, we have for each l

$$\|U(b_l) - I(h_l)\| = \|\Sigma_{n_{l-1}<i\leq n_l} I\ T_i(h_l) - I(h_l)\| \leq$$

$$\|\Sigma_{i\leq n_{l-1}} I\ T_i(h_l)\| + \|\Sigma_{i>n_l} U\ P_i\ I\ (h_l)\| <$$

$$\varepsilon_l(\|I\| + \|U\|)\cdot \|h_l\|_p < 4^{-l}\ \|I^{-1}\|^{-1}\ \|h_l\|_p$$

and also

$$\| V I^{-1} U(b_l) - h_l \| < 4^{-1} \| h_l \|_p .$$

From these estimates, it is not difficult to see that (h_l), $(I h_l)$, $(U b_l)$ and $(V I^{-1} U b_l)$ are equivalent sequences and moreover $[h_l] = [V I^{-1} U (b_l)]^{(1)}$.

Let λ be the unconditionality constant of the decomposition (X_i). For a given finite sequence (a_l) of scalars, there is a finite sequence (b_l^*) in X^* such that

(iv) $b_l^* \in [P_i^* X_i^*]_{i=n_{l-1}+1}^{n_l}$

(v) $\| \Sigma_l \varepsilon_l b_l^* \| \leq 2\lambda$ for all signs $\varepsilon_l = \pm 1$

(vi) $\Sigma_l a_l \langle b_l, b_l^* \rangle = \| \Sigma_l a_l b_l \|$.

Since also $(I h_l)$ is unconditional of constant c only depending on p and $\|I\| \, \|I^{-1}\|$, we get

$$\| \Sigma_l a_l I(h_l) \| \geq c^{-1} \int_0^1 \| \Sigma_l a_l r_l(\omega) I(h_l) \| d\omega$$

$$\geq \frac{1}{2\lambda c} \int \langle \Sigma_l a_l r_l(\omega) I(h_l), \Sigma_l r_l(\omega) \varepsilon_l b_l^* \rangle d\omega$$

$$= \frac{1}{2\lambda c} \Sigma_l |a_l| \, |\langle I(h_l), b_l^* \rangle|$$

for the right choice of signs $\varepsilon_l = \pm 1$.

Because $\langle I(h_l), b_l^* \rangle = \langle \Sigma_{i=n_{l-1}+1}^{n_l} P_i I (h_l), b_l^* \rangle = \langle b_l, b_l^* \rangle$

we conclude that $\| \Sigma_l a_l b_l \| \leq 2\lambda c \| \Sigma_l a_l I(h_l) \|$.

This, combined with (1), implies that (b_l) and (h_l) are equivalent. Call $W : [h_l] \to [b_l]$ the isomorphism assigning b_l to $V I^{-1} U (b_l)$. It is easily verified that $[b_l]$ is complemented in X by $W V I^{-1} U$.
This ends the proof of 4.36

12. The reader will find additional results concerning $\mathcal{L}^p$-subspaces of Y_p in [ ]..

V. A CLASS OF $\mathcal{L}^1$-SPACES

1. INTRODUCTION

The characterization of the complemented subspaces of L^1 is an open problem. It is conjectured that such a space is either isomorphic to l^1 or to L^1. The answer turns out to be affirmative in case of a norm-1 projection (see [87]). It is also known that a complemented subspace of L^1 which has the Radon-Nikodým property is isomorphic to l^1.

One may hope for relations between the failure of the Radon-Nikodým property, the failure of the Schur property and the imbedding of L^1 for complemented subspace of L^1. The only results in this context is a theorem due to H. Rosenthal (see [17]) claiming that a non-Schur complemented subspace of L^1 has an l^2-subspace and an improvement of the author (see [17]) who obtained an $\oplus_{l^1}(l^2)$-subspace.

It was shown in the first chapter that complemented subspaces of L^1 are $\mathcal{L}^1$-spaces. The converse property however is wrong. For the proof of the following two results, we refer to [87] .

<u>PROPOSITION 5.1</u> : Let X be a separable $\mathcal{L}^1$-space and $T : l^1 \to X$ a quotient map. Then the kernel of T is also a $\mathcal{L}^1$-space.

<u>PROPOSITION 5.2</u> : Assume Y_1 and Y_2 isomorphic infinite-dimensional subspaces of l^1 so that $X_1 = l^1/Y_1$ and $X_2 = l^2/Y_2$ are infinite-dimensional $\mathcal{L}^1$-spaces. Then X_1 and X_2 are isomorphic.

We may introduce a sequence of subspaces of l^1 as follows :
Take D_0 such that l^1/D_0 is isomorphic to L^1.
If D_k is obtained, define D_{k+1} such that l^1/D_{k+1} is isomorphic to D_k.

It is clear from (5.1) that the D_k are $\mathcal{L}^1$-spaces. Applying (5.2), we see that they are moreover mutually non-isomorphic.
Since the D_k have the RNP and are not isomorphic to l^1, it follows that they are not isomorphic to a complemented subspace of L^1.

The problem about the existence of uncountably many separable $\mathcal{L}^1$-types was solved positively by joint work of W.B. Johnson and

J. Lindenstrauss (see [70]) and is based on some ideas due to P.W. Mc Cartney and R.C. O'Brien (see [91]). We sketch briefly the construction.

Let $T : l^1 \to L^1$ be a fixed quotient map and define for each $0 < \alpha < 1$ X_α as the subspace $\{(\alpha x, Tx) \; ; \; x \in l^1\}$ of $(l^1 \oplus L^1)_1$. It can be shown that X_α (which is obviously isomorphic to l^1) is a $\mathcal{L}^1_\lambda$-space, where λ does not depend on α in particular. For a decreasing sequence $1 > \alpha_1 > \alpha_2 > \ldots$ tending to 0, the l^1-sum $Y = Y(\alpha_n \; ; \; n)$ is still a $\mathcal{L}^1_\lambda$-space, satisfying the Schur property and the RNP. Now Y has the so called neighborly-tree-property and therefore can not be embedded in a separable dual space. Thus this gives another counterexample to the UHL-conjecture mentioned in Ch. III. It turns moreover out that, by suitable choices of the sequence (α_n), the procedure described above gives a continuum number of mutually non-isomorphic $\mathcal{L}^1$-spaces. However, all these spaces Y can be embedded in the l^1-sum of the spaces X_α where $0 < \alpha < 1$ and α is rational and the latter space is still Schur and Radon-Nikodým.

The aim of this chapter is to prove the following fact, solving a question raised by A. Pelczynski (see [30]).

<u>THEOREM 5.3</u> : The class of separable $\mathcal{L}^1$-spaces not containing a copy of L^1 has no universal element.

Examples will be given of $\mathcal{L}^1$-spaces with the RNP failing the Schur property and vice versa.

2. A CONSTRUCTION TECHNIQUE FOR $\mathcal{L}^1$-SPACES

The purpose of this section is to prove the following result, which is the starting point in the construction of our $\mathcal{L}^1$-spaces.

<u>THEOREM 5.4</u> : Assume T an operator on L^1 and E a subspace of L^1 such that the restriction of T to E is the identity.

We consider the following properties for T

(a) T does not induce an isomorphism on an L^1-subspace

(b) For operators S from an $L^1(\mu)$-space into L^1, the representability of $(I-T)S$ implies the representability of S.

(c) A weakly compact subset of L^1 is norm-compact, provided its image by $I-T$ is compact.

In the respective cases (a), (b), (c), the space E embeds in a $\mathcal{L}^1$-space B satisfying the corresponding property (or properties)

(a') L^1 does not embed in B

(b') B has the RNP

(c') B has the Schur property.

Proof : It is rather simple. Fixing $\rho > 1$, one can find a sequence of subspaces U_i of L^1 satisfying the following conditions

1. Each U_i is finite dimensional, let us say $d_i = \dim U_i$
2. $d(U_i, \ell^1(\dim U_i))$
3. $U_i \subset U_{i+1}$
4. $T(U_i) \subset U_{i+1}$
5. $\bigcup_{i=1}^{\infty} U_i$ is dense in L^1
6. $\bigcup_{i=1}^{\infty} (E \cap U_i)$ is dense in E

That this can be done is straightforward and we let the reader check the details.

In what follows, $\oplus$ will denote the direct sum in ℓ^1-sense.

Define

$$\mathfrak{X} = L^1 \oplus \oplus_{i=1}^{\infty} U_i$$

and let $P : \mathfrak{X} \to L^1$ and $P_i : \mathfrak{X} \to U_i$ be the projections.

We further introduce for each j the space

$$_j = U_j \oplus \oplus_{i=1}^{j} U_i$$

which embeds in $\mathfrak{X}$ in natural way.

For fixed j, let $I_j : \mathfrak{X}_j \to \mathfrak{X}$ be the operator defined as follows

$$\begin{cases} P\, I_j(x) = T\, P(x) \\ P_i\, I_j(x) = P_i(x) \quad \text{for } i=1,\dots,j \\ P_{j+1}\, I_j(x) = P(x) - T\,P(x) - \Sigma_{i\leq j}\, P_i(x) \\ P_i\, I_j(x) = 0 \quad \text{for } i > j+1 \end{cases}$$

which makes sense by conditions (3) and (4) on the spaces U_i.
Remark that

$$P(x) = (P + \Sigma_i\, P_i)\, I_j(x) \qquad (*)$$

for all $x \in \mathfrak{X}_j$.

Since the following inequalities are clearly satisfied

$$\tfrac{1}{2}\|x\| \leq \|I_j(x)\| \leq 2(1 + \|T\|)\|x\|$$

we see that I_j is an isomorphism on its range B_j.
More precisely, we have
$d(B_j, \mathfrak{X}_j) \leq 4(1 + \|T\|)$
and thus
$d(B_j, \ell^1(d_j)) \leq 4\,\rho(1 + \|T\|)$

Our next claim is that B_j is a subspace of B_{j+1}. Let indeed $x \in \mathfrak{X}_j$ and define y by

$$\begin{cases} P(y) = P(x) \\ P_i(y) = P_i(x) \quad \text{for } i=1,\dots,j \\ P_{j+1}(y) = P(x) - T\,P(x) - \Sigma_{i\leq j}\, P_i(x) \\ P_i(y) = 0 \quad \text{for } i > j+1 \end{cases}$$

which is a member of $\mathfrak{X}_{j+1}$. A simple verification shows that
$I_{j+1}(y) = I_j(x)$.
This shows that $B_j \subset B_{j+1}$.

L^1 embeds in $\mathfrak{X}$ by identification with the first coordinate.
By hypothesis on T, one has that $I_j(x) = x$ whenever
$x \in E \cap U_j \hookrightarrow \mathfrak{X}_j$. Thus $E \cap U_j$ is a subspace of B_j and we conclude that E is a subspace of B.

We show that L^1 does not embed in B if T does not fix an L^1-copy. As a consequence of (*), we get

$$P(x) = T(P + \Sigma_i P_i)(x) \quad \text{for all} \quad x \in B \qquad (**)$$

leading to the following scheme

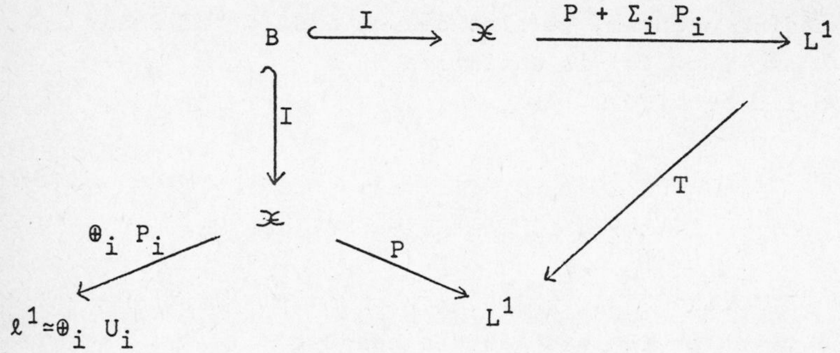

If B contains an L^1-subspace, one of the operators $\oplus_i P_i \circ I$, $P \circ I$ has to fix a copy of L^1 (see [56]). Since $\oplus_i P_i \circ I$ ranges in $\oplus_i U_i$ which is isomorphic to ℓ^1 and $P \circ I$ factors over T, T will fix an L^1-space. This proves (a) of the Theorem.

In order to show (b), we rewrite (**) in the form

$$(I-T)P\,(x) = T(\Sigma_i P_i)(x) \quad \text{for all} \quad x \in B \qquad (***)$$

which gives the diagram

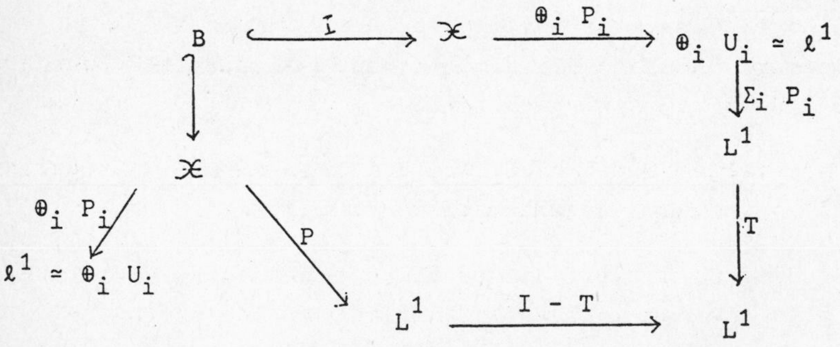

Suppose now $S : L^1(\mu) \to B$ a non-representable operator. Since $S = PS \oplus (\oplus_i P_i)S$ and $\oplus_i P_i$ ranges in ℓ^1, it follows that PS will be non-representable. But, under hypothesis (b) of the theorem, this implies the non-representability of the operator (I-T)PS factoring through ℓ^1, a contradiction.

Using the preceding diagram, we also deduce easily the Schur property for B provided (c) is satisfied.
This completes the proof of 5.4.

The main point in the application of 5.4 is of course the construction of operators on L^1 and the verification of properties like (a), (b), (c).
We will first establish 5.3. The construction of "small" non-RN $\mathcal{L}^1$ spaces will be given at the end of this chapter.

In fact, Th. 5.3 will extend the preceding chapter. The starting point are the tree-spaces X_T^p introduced in this chapter (see section 7). We know that they are complemented in $L^p(G)$ for $1 < p < \infty$. This is obviously not the case for $p = 1$ but we will show that the X_T^1-spaces for T well-founded can be obtained as proper subspace of a convolution operator on G satisfying (b). Application of 5.4 will then allow to embed X_T^1 in a $\mathcal{L}^1$-space with the RNP.

3. CONSTRUCTION OF CERTAIN MEASURES ON G

Denote by C(G) the space of continuous functions on G.
If μ is a measure on G and S a finite subset of C, the S-Fourier coefficient $\hat{\mu}(S)$ of μ is given by $\hat{\mu}(S) = \int_G w_S \, d\mu$.

If μ and ν are two measures on G, the convolution $\mu * \nu$ is defined by $(\mu * \nu)(f) = \int f(x.y)\, \mu(dx)\, \nu(dy)$ for $f \in C(G)$.

For a subset S of C, $\mathcal{E}_S$ will denote the conditional expectation with respect to the sub-σ-algebra generated by the Rademacker functions $(r_c)_{c \in S}$.

If μ is a measure on G and S a subset of C, we say that μ is S-dependent provided $\mu = \mu \circ \mathcal{E}_S$.

For S subset of C, take $\tilde{S} = \{c \in C \; ; \; c < d$ for some $d \in S\}$.
For convenience, we introduce the "empty complex" (cfr. [42]), denoted by the symbol ϕ. Define $C^* = C \cup \{\phi\}$.
If $c \in C$, let $c' \in C^*$ be the predecessor of c.

A weightfunction will be a function τ on C^* ranging in the open interval $]0,1[$ such that $K_\tau = \prod_{c \in R^*} (2\,\tau(c)^{-1} - 1) < \infty$.
If τ is a weightfunction, take $[\tau \; , \phi] = 1$ and $[\tau \; ; S] = \prod_{c \in \tilde{S}} \tau(c')$, if S is a nonempty finite subset of C.

The next lemma is needed for later purpose.

<u>LEMMA 5.5</u> : Let τ be a weightfunction, $\varepsilon > 0$ and define

$$\mathcal{S}_\varepsilon = \{S \subset C \; ; \; S \text{ finite and } [\tau \; ; S] > \varepsilon\}$$

Assume further (S_k) a sequence of disjoint finite subsets of C so that for each k one can find $S \in \mathcal{S}_\varepsilon$ with $S \cap S_\ell \neq \phi$ for each $\ell = 1,\ldots,k$. Then there exists a sequence (n_i) of integers such that $(n_1,\ldots,n_i) \in \cup_k \tilde{S}_k$ for each j.

<u>Proof</u> : Let us first remark that for fixed $c \in C^*$ we find $[\tau \; ; S] \leq \tau(c)^{|M|}$ for each $S \subset C$, where $M = \{n \in \mathbb{N} \; , \; (c,n) < d$ for some $d \in S\}$.

Because $\tau(c) < 1$, this implies that $|M|$ is uniformely bounded for S ranging in $\mathcal{S}_\varepsilon$.
Since the S_k are finite sets, one can pick elements $c_k \in S_k$ such that for all k there is some $S \in \mathcal{S}_\varepsilon$ with $\{c_1, c_2, \ldots, c_k\} \subset S$.
It is now possible to construct a sequence of integers (n_i), so that for each j the set $\{k \; ; \; (n_1,\ldots,n_j) < c_k\}$ is infinite.
We indicate briefly the inductive procedure. Suppose $n_1,\ldots,n_j$ obtained. By hypothesis, any finite subset of the set $\{(n_1,\ldots,n_j,n) \; ; \; n \in N\}$, where $N = \{n \in \mathbb{N} \; ; \; (n_1,\ldots,n_j,n) < c_k$, for some k$\}$, is contained in $\tilde{S}$ for some $S \in \mathcal{S}_\varepsilon$. By the first

observation we made, we conclude that N is finite. This allows us to choose n_{j+1} so that $\{k \; ; \; (n_1,\dots,n_j,n_{j+1}) < c_k\}$ is again infinite.
Clearly, for each j, $(n_1,n_2,\dots,n_j) \in \tilde{S}_k$ for some k and this ends the proof.

The main objective of this section is to prove the following result

PROPOSITION 5.6 : Let τ be a weightfunction. Then there exists a measure μ on G satisfying the following properties

1. $\|\mu\| \leq K_\tau$
2. $\hat{\mu}(S) = 1$ if S is a branch set
3. If S is a finite subset of $\mathcal{C}$ and $f \in C(G)$ is $(\mathcal{C}\setminus S)$-dependent then $|\int w_S(x)\, f(x)\, \mu(dx)| \leq K_\tau[\tau \; ; \; S]\, \|f\|_\infty$.

Proof : For $c \in \mathcal{C}^*$, let
$\mathcal{C}_c^* = \{d \in \mathcal{C}^* \; , \; c < d\}$ and $\mathcal{C}_c = \{d \in \mathcal{C} \; , \; c < d \text{ and } c \neq d\}$.
We let δ be the Dirac-measure on G and δ_c the Dirac-measure on the c-factor $\{1,-1\}$ in the product G.

If for fixed $c \in \mathcal{C}$, we define the measure ν_c on G by

$$\nu_c = \tau(c')\,\delta + (1 - \tau(c'))\,\Big(\bigotimes_{d\in\mathcal{C}_c^*} m_d \otimes \bigotimes_{d\in\mathcal{C}\setminus\mathcal{C}_c^*} \delta_d\Big)$$

then clearly

4. $\|\nu_c\| = 1$
5. $\hat{\nu}_c(S) = 1$ if $c \notin \tilde{S}$ and $\hat{\nu}_c(S) = \tau(c')$ if $c \in \tilde{S}$.
6. If S is a finite subset of $\mathcal{C}$ and $f \in C(G)$ is $(\mathcal{C}\setminus S)$-dependent, then $|\int w_S(x)\, f(x)\, \nu_c(dx)| \leq \hat{\nu}_c(S)\, \|f\|_\infty$.

Let ν be the measure on G obtained by convolution of the ν_c, thus

$$\nu = \mathop{*}_{c\in\mathcal{C}} \nu_c$$

As an easy verification shows, the following holds

7. $\|\nu\| = 1$
8. $\hat{\nu}(S) = [\tau \; ; \; S]$

9. If S is a finite subset of C can $f \in C(G)$ is $(C \setminus S)$-dependent, then $|\int w_S(x)\ f(x)\ \nu(dx)| \leq [\tau\ ;\ S] \|f\|_\infty$.

Next, consider for fixed $c \in C^*$ the following measure η_c on G

$$\eta_c = \tau(c)^{-1}\ \delta + (1 - \tau(c)^{-1})(\bigotimes_{d \in C_c} m_d \otimes \bigotimes_{d \in C \setminus C_c} \delta_d)$$

satisfying

10. $\|\eta_c\| \leq 2\tau(c)^{-1} - 1$
11. $\hat{\eta}_c(S) = 1$ if $C_c \cap S = \phi$ and $\hat{\eta}_c(S) = \tau(c)^{-1}$ if $C_c \cap S \neq \phi$.

Since τ is a weightfunction, we can define the convolution η of the η_c, thus

$$\eta = \mathop{\ast}_{c \in C^*} \eta_c$$

for which

12. $\|\eta\| \leq K_\tau$

Finally, take $\mu = \nu \ast \eta$. Then clearly $\|\mu\| \leq \|\nu\|\ \|\eta\| \leq K_\tau$.

In order to verify (2), let S be a finite branch set. Then $\hat{\mu}(S) = \hat{\nu}(S).\hat{\eta}(S) = [\tau\ ;\ S]\ .\ \Pi\{\tau(c)^{-1}\ ,\ c \in C^*$ with $C_c \cap S \neq \phi\} = [\tau\ ;\ S]\ .\ \Pi_{c \in \tilde{S}}\ \tau(c')^{-1} = 1$, as required.

Let us check (3). So take S a finite subset of C and an $(C \setminus S)$-dependent function $f \in C(G)$. We have

$$\int w_S(x)\ f(x)\ \mu(dx) = \int w_S(x.y)\ f(x.y)\ \nu(dx)\ \eta(dy)$$

and thus

$$\begin{aligned} |\int w_S(x)\ f(x)\ \mu(dx)| &\leq \int |\int w_S(x)\ f(x.y)\ \nu(dx)|\ |\eta|\ (dy) \\ &\leq \|\eta\|\ [\tau\ ;\ S]\ \sup_y \|f_y\|_\infty \\ &\leq K_\tau\ [\tau\ ;\ S]\ \|f\|_\infty \end{aligned}$$

completing the proof.

5. NON-REPRESENTABLE OPERATORS RANGING IN L^1-PRODUCTSPACES

Our purpose is to present here some technical ingredients needed for the next section.

The following result is essentially known, but we include its proof here for selfcontainedness.

LEMMA 5.7 : Let μ, ν be probability spaces and $S : L^1(\mu) \to L^1(\nu)$ a non-representable operator. Then there exist a bounded convex subset C of $L^1(\nu)$ and $\rho > 0$ such that the following holds whenever $T : L^1(\nu) \to B$ is an operator with TS representable :
If $f \in C$ and $\delta > 0$, then there exists some $g \in C$ so that $\|T(f-g)\| < \delta$ and $\int_A |g| \, d\nu \geq \rho$ for some ν-measurable set A with $\nu(A) < \delta$.

Proof : If S is not representable, then one can find a μ-measurable set Ω with $\mu(\Omega) > 0$ and $\rho > 0$ such that for any $\Omega' \subset \Omega$, $\mu(\Omega') > 0$ and $\delta > 0$, there exists $\Omega'' \subset \Omega'$, $\mu(\Omega'') > 0$ with $\int_A |S(\Omega'')| \, d\nu > \rho \, \mu(\Omega'')$ for some A with $\nu(A) < \delta$ (cfr. [45]).

If now $\Omega_1, \ldots, \Omega_d$ are subsets of Ω with positive measure and $\delta > 0$, there exist subsets $\Omega_i' \subset \Omega_i$ $(1 \leq i \leq d)$ of positive measure satisfying the following condition :
There exists a set A with $\nu(A) < \delta$ such that

$$\int_A |\Sigma_i \, a_i \, S(\Omega_i')| \, d\nu \geq \rho \, \Sigma_i \, |a_i| \, \mu(\Omega_i')$$

for all scalars $a_1, \ldots, a_d$.
The proof of the latter fact is elementary and left as an exercice to the reader.

We show that

$$C = \{S(\varphi) \; ; \varphi \in L^1_+(\Omega), \int \varphi \, d\mu = 1\}$$

satisfies the condition of the lemma.

So fix $f = S(\varphi)$, $\varphi \in L^1_+(\Omega)$, $\int \varphi \, d\mu = 1$ and $\delta > 0$.

Using the representability of the operator TS, it is possible to find a partition $\Omega_1,\ldots,\Omega_d$ of Ω and scalars $a_1,\ldots,a_d$, such that

(i) $\left\| \frac{TS(\Omega_i)}{\mu(\Omega_i)} - \frac{TS(\Omega')}{\mu(\Omega')} \right\| < \tau$, whenever $\Omega' \subset \Omega_i$, $\mu(\Omega') > 0$

(ii) $\|\varphi - \Sigma_i\ a_i\ \chi_{\Omega_i}\|_1 < \tau$

(iii) $\Sigma_i\ a_i\ \mu(\Omega_i) = 1$

where $\tau = \frac{\delta}{1 + \|TS\|}$.

By the previous observation, one can obtain subsets $\Omega_i' \subset \Omega_i$ $(1 \leqslant i \leqslant d)$ of positive measure and a set A with $\nu(A) < \delta$, so that

$$\int_A |\Sigma_i\ a_i\ \frac{\mu(\Omega_i)}{\mu(\Omega_i')}\ S(\Omega_i')|\ d\nu \geqslant \rho\ \Sigma_i\ a_i\ \mu(\Omega_i) = \rho.$$

Take $\psi = \Sigma_i\ a_i\ \frac{\mu(\Omega_i)}{\mu(\Omega_i')}\ \chi_{\Omega_i'}$ and $g = S(\psi)$, belonging to C.

Thus $\int_A |g|\ d\nu \geqslant \rho$ and $\|T(f-g)\| = \|TS(\varphi) - TS(\psi)\| \leqslant$

$$\tau\|TS\| + \Sigma_i\ a_i\ \|TS(\Omega_i) - \frac{\mu(\Omega_i)}{\mu(\Omega_i')}\ TS(\Omega_i')\| < (1 + \|TS\|)\tau,$$

completing the proof.

<u>COROLLARY 5.8</u> : Under the hypothesis of Lemma 5.7, one can find $C \subset L^1(\nu)$ and $\rho > 0$ such that whenever $T : L^1(\nu) \to L^1(\nu)$ is an operator with TS representable, $f \in C$ and $\delta > c$, there exists $g \in C$ satisfying $\|f-g\| > \rho$, $|\int (f-g)d\nu| < \delta$ and $\|T(f-g)\| < \delta$.

Assume now $(\Omega_i,\nu_i)_{i\in D}$ a family of probability spaces and consider the product space $(\Omega,\nu) = (\Pi_i\ \Omega_i,\ \otimes_i\ \nu_i)$. For $E \subset D$, denote $\mathcal{E}_E$ the corresponding conditional expectation.
We claim the following

LEMMA 5.9 : If $S : L^1(\mu) \to L^1(\nu)$ is non-representable, then D has a finite subset E so that $(I - \mathcal{E}_{D\setminus E}).S$ is non-representable too.

Proof : Let $C \subset L^1(\nu)$ and $\rho > 0$ be as in the preceding corollary. Assume the above statement wrong. Successive applications of 5.8 allow us then to construct a sequence (f_k) in C and an increasing sequence (E_k) of finite subsets of D in such a way that

1. $\| f_k - \mathcal{E}_{E_k}[f_k] \|_1 < 2^{-k}$

2. $\| f_k - f_{k+1} \|_1 > \rho$

3. $|\int (f_k - f_{k+1}) d\nu| < 2^{-k}$

4. $\| (f_k - f_{k+1}) - \mathcal{E}_{D\setminus E_k}[f_k - f_{k+1}] \|_1 < 2^{-k}$

Take then $\varphi_1 = \mathcal{E}_{E_1}[f_1]$ and $\varphi_k = \mathcal{E}_{E_k \setminus E_{k-1}}[f_k - f_{k-1}]$ for $k > 1$.

By (1) and (4), we find that $\| f_k - f_{k-1} - \varphi_k \|_1 < 8.2^{-k}$.

Consequently, by (2), $\|\varphi_k\|_1 > \rho - 8.2^{-k}$ and, by (3), $|\int \varphi_k \, d\nu| < 10.2^{-k}$.

Consider now the sequence $\psi_k = \varphi_k - \int \varphi_k \, d\nu$, which consists of independent mean-zero functions. Since (ψ_k) is an unconditional basic sequence in $L^1(\nu)$, (ψ_k) is also boundedly complete. But $\|\psi_k\| \geq \rho - 18\ 2^{-k}$ and on the other hand

$\| \Sigma_{k=1}^n \psi_k \|_1 \leq 2 \| \Sigma_{k=1}^n \varphi_k \|_1 \leq 2 \| f_n \|_1 + 8\ \Sigma_{k=1}^n 2^{-k} < \sup \{ \|f\|_1 \ ;\ f \in C \} + 8$, a contradiction.

Repeating applications of lemma 5.9 leads to the next

COROLLARY 5.10 : Suppose moreover (Ω_i, ν_i) purely atomic for each $i \in D$. Then, under the hypothesis of Lemma 9, there is a sequence (E_k) of disjoint finite subsets of D, so that for all k the operator $\Pi_{\ell=1}^k (I - \mathcal{E}_{D\setminus E_\ell}) \circ S$ is not representable.

6. APPLICATION TO CERTAIN OPERATORS ON $L^1(G)$

Referring to section 4, let τ be a fixed weightfunction and let μ be the measure on G constructed in proposition 5.6.
Consider the operator Λ on $L^1(G)$ obtained by μ-convolution, i.e.

$$\Lambda(f)(x) = \int_G f(x.y)\ \mu(dy)$$

The following is easily derived from proposition 5.6.

PROPOSITION 5.11 :

1. $\|\Lambda\| \leqslant K_\tau$
2. $\Lambda(w_S) = w_S$ if S is a branch
3. If S is a finite subset of $\mathcal{R}$ and $f \in L^1(G)$ is S'-dependent, where $S \cap S' = \phi$, then $\Lambda(w_S \otimes f) = w_S \otimes \overline{f}$ for some S'-dependent function $\overline{f}$ in $L^1(G)$ satisfying $\|\overline{f}\|_1 \leqslant K_\tau[\tau\ ;\ S]\ \|f\|_1$.

For any well-founded tree T on $\mathbb{N}$, define the operator Λ_T on $L^1(G)$ by $\Lambda_T = \mathcal{E}_T \circ \Lambda = \Lambda \circ \mathcal{E}_T$.

It is clear from Proposition 5.11 (2) that Λ_T is the identity on X_T^1.

Finally, it remains to establish

PROPOSITION 5.12 : If $\Gamma : L^1(\lambda) \to L^1(G)$ is a non-representable operator, then $(I - \Lambda_T) \circ \Gamma$ is non-representable to, for any well-founded tree T.

Since $G = \{1,-1\}^{\mathcal{C}}$, application of corollary 5.10 yields a sequence (S_k) of disjoint finite subsets of $\mathcal{C}$, so that for all k the operator $\Phi_k \circ \Gamma$ is non-representable, where $\Phi_k = \Pi_{\ell=1}^k (I - \mathcal{E}_{\mathcal{C}\setminus S_\ell})$.

Assume $(I - \Lambda_T) \circ \Gamma$ representable and take $\varepsilon = \frac{1}{2K_\tau}$. Then

LEMMA 5.13 : For each k, there is some $S \in \mathcal{S}_\varepsilon$ such that $S \subset T$ and $S \cap S_\ell \neq \phi$ for each $\ell = 1,\dots,k$.

Once this obtained, we may apply lemma 5.5, taking the sequence $(S_k \cap T)$ in account. This leads to a sequence (n_i) of integers with $(n_1,n_2,\dots,n_j) \in T$ for each j, contradicting the assumption that T was well-founded. So it remains to prove the above lemma.

Proof of Lemma 5.13 : Fix k, consider the set
$\mathcal{F} = \{S \subset T \; ; \; S \subset \cup_{\ell=1}^{k} S_\ell$ and $S \cap S_\ell \neq \phi$ for each $\ell = 1,\ldots,k\}$
and denote r its cardinality.
Since $\Phi_k \circ \Gamma$ is non-representable and, by hypothesis,
$\Phi_k \circ (I - \Lambda_T) \circ \Gamma = (I - \Lambda_T) \circ \Phi_k \circ \Gamma$ is representable, the operators $\Lambda_T \Phi_k \Gamma$ and hence $\mathcal{E}_T \Phi_k \Gamma$ are non-representable.
Therefore, there exists some $\varphi \in L^1(\lambda)$ satisfying
$\|\mathcal{E}_T \Phi_k \Gamma(\varphi)\| = 1$ and $\|(I-\Lambda_T) \Phi_k \Gamma(\varphi)\| < \frac{1}{3r}$.
Define $f = \mathcal{E}_T \Phi_k \Gamma(\varphi)$, which is clearly of the following form

$$f = \Sigma_{S\in\mathcal{F}} w_S \otimes f_S$$

where each f_S is $(T\setminus\cup_{\ell=1}^{k} S_\ell)$-dependent.

Moreover, by construction, $\|f\| = 1$ and $\|(I-\Lambda)f\| < \frac{1}{3r}$.

By proposition 5.11 (3), we see that $\Lambda(w_S \otimes f_S) = w_S \otimes \overline{f}_S$ for some $(T\setminus\cup_{\ell=1}^{k} S_\ell)$-dependent function $\overline{f}_S$ in $L^1(G)$ satisfying
$\|\overline{f}_S\|_1 \leq K_\tau[\tau \; ; \; S] \; \|f_S\|_1$. Thus

$$\Lambda(f) = \Sigma_{S\in\mathcal{F}} w_S \otimes \overline{f}_S$$

and

$$f-\Lambda(f) = \Sigma_{S\in\mathcal{F}} w_S \otimes (f_S-\overline{f}_S)$$

For each $S \in \mathcal{F}$, we have
$\frac{1}{3r} > \|f-\Lambda(f)\|_1 \geq \|f_S-\overline{f}_S\|_1 \geq \|f_S\|_1 - \|\overline{f}_S\|_1 \geq (1 - K_\tau[\tau \; ; \; S])\|f_S\|_1$
and hence for $S \in \mathcal{F}\setminus\mathcal{S}_\varepsilon$, by the choice of ε
$\|f_S\|_1 < \frac{2}{3r}$.

Suppose $\mathcal{F} \cap \mathcal{S}_\varepsilon = \phi$. Then it would follow
$1 = \|f\|_1 \leq \Sigma_{S\in\mathcal{F}} \|f_S\|_1 < \frac{2|\mathcal{F}|}{3r}$,
a contradiction.

Consequently, $\mathcal{F} \cap \mathcal{S}_\varepsilon \neq \phi$, completing the proof.

Proof of Theorem 5.3 : By 4.37, it suffices to embed any tree-space X_T^1, for T well-founded, in an RN $\mathcal{L}^1$-space. Now Λ_T is an operator on $L^1(G)$ which is identity on X_T^1 and satisfies 5.12. So we can apply 5.4 (b).

We conclude this section with some remarks and questions.

1. One can show that the operators Λ_T for T well-founded do not fix an L^1-copy and hence also satisfy condition (a) of theorem 1.

2. As far as we know the $\mathcal{L}^1$-spaces constructed here are also the first examples of non-Schur $\mathcal{L}^1$-spaces which do not contain a copy of L^1.

3. Related to this work and also [15] is the following question : Does the class of separable Schur $\mathcal{L}^1$-spaces admit an universal element ? and its weaker version.
Does there exist a separable Banach space not containing L^1 which is universal for latter class of spaces ?

6. $\mathcal{L}^1$-SPACES WITH THE SCHUR PROPERTY FAILING THE RNP

As another application of the general construction principle explained in section 2, we will show the existence of "small" $\mathcal{L}^1$-spaces failing the RNP. Let us point out that the problem whether or not a non-RNP-subspace of L^1 has to contain an L^1-copy isomorphically remained open for some time and was first solved (negatively) in [31].

The building pieces of the required operator on L^1 will again be convolution operators. Let us first introduce some notation.
For each positive integer N, let G_N be the Cantor group $\{1,-1\}^N$ and m_N the Haar-measure of G_N. For $1 \leq n \leq N$, the nth Rademacher function r_n on G_N is defined by $r_n(x) = x_n$. For $0 \leq \varepsilon \leq 1$ fixed, T_ε is the convolution operator on $L^1(G_N)$ by the measure
$\prod_n (1+\varepsilon r_n)$
Thus $T_\varepsilon(w_S) = \varepsilon^{(S)} w_S$ for Walshes $w_S = \prod_{n\in S} r_n$, taking $|S|$ the cardinality of S.
Let r be a fixed positive integer. We consider the probability space (Ω_r,μ_r) obtained as direct sum of G_N' and G_N'' where

$N = N_r = r^{(8^r)}$ and where G_N' and G_N'' are the group G_N equipped with the respective measures $(1-2^{-r})m_N$ and $2^{-r}m_N$. Take $\varepsilon = 1/r$. For $\nu \in G_N$, we introduce following functions on G_N

$$e_\nu = \prod_n (1+\nu_n r_n) \quad \text{and} \quad f_\nu = \frac{1}{d} \sum_{i=0}^{d-1} T_{\varepsilon^i}(e_\nu) \quad \text{where} \quad d = 4^r.$$

Finally, define as follows an operator T_r on $L^1(\mu_i)$:

$$T_r(f' \oplus f'') = (T_\varepsilon f' + \frac{1}{d}(f'' - T_{\varepsilon^d} f'')) \oplus f''$$

In what follows, we will use following facts about T_r.

<u>LEMMA 5.14</u> : If for each $\nu \in G_N$ we consider the function $\varphi_\nu = f_\nu \oplus e_\nu$ on Ω_r, then

1. $\|\varphi_\nu\|_1 = \int \varphi_\nu \, d\mu_i = 1$
2. $2^{-N} \sum \varphi_\nu$ is the constant function 1 on Ω_r
3. $\|\varphi_\nu - 1\|_1 > 2 - \frac{1}{r}$
4. $\|T_r\| \leq 1 + 2^{-r+1}$
5. $T_r(\varphi_\nu) = \varphi_\nu$

The verification of assertions (1), (2), (4), (5) is almost immediate.

Property (3) follows from the fact that by construction φ_ν is localized on a set of small measure. We check this using standard techniques, for instance by estimation of $\|\sqrt{\varphi}\|_1$.

Next, define $\Omega = \prod_n \Omega_r$ equipped with product measure $\mu = \otimes_r \mu_r$.

Since $\prod \|T_r\| < \infty$, we are allowed to define $T = \otimes_r T_r$ as the product-operator on $L^1(\mu)$.

Let E be the subspace of $L^1(\mu)$ generated by the functions

$$\varphi_{\nu^1} \otimes \varphi_{\nu^2} \otimes \ldots \otimes \varphi_{\nu^r}$$

where

$$r = 1,2,\dots \quad \text{and} \quad \nu^1 \in G_{N_1},\ \nu^2 \in G_{N_2},\ \dots,\ \nu^r \in G_{N_r}$$

It follows from (1), (2), (3) of 5.14 that this functions form a bounded and nowhere convergent tree in $L^1(\mu)$. Hence E fails RNP. We deduce from (5) that T restricted to E is identity.

By 5.4, in order to embed E in a Schur $\mathcal{L}^1$-space, it suffices to establish following

LEMMA 5.15 : A weakly compact subset of $L^1(\mu)$ is norm-compact, provided its image by I - T is compact.

At this point, the fact that $\varepsilon_r \to 0$ for $r \to \infty$ will be of importance. We first verify

LEMMA 5.16 : Denote $\mathcal{E}_r$ the expectation with respect to $\bigotimes_{s \leq r} \Omega_s$. If $F \in L^\infty(\mu)$ and $\mathcal{E}_r[F] = 0$, then

$$\|T(F)\|_1 \leq 3/r\ \|T\|\ \|F\|_\infty.$$

Proof : It is in similar spirit as 5.13. Introduce the subset

$$\Phi = \prod_{s \leq r} \Omega_s \times \prod_{s > r} G'_{N_s}$$

of Ω and the restriction G of F to Φ.
From the definition of the measures μ_s, we find

$$\|F-G\|_2 \leq \|F\|_\infty\ \mu(\Omega \setminus \Phi)^{1/2} \leq 2^{-r/2}\ \|F\|_\infty$$

and hence

$$\|T(F)\|_1 \leq \|T(G)\|_1 + 1/r\ \|T\|\ \|F\|_\infty.$$

Take $U = \bigotimes_{s \leq r} T_s$ and $V = \bigotimes_{s > r} T_s$. Rewrite G as Walsh-expansion

$$G = \sum_S G_S \otimes W'_S$$

where S runs over the sequences $(S_{i+1}, S_{i+2}, \dots)$ of subsets S_s of

$\{1,2,\ldots,N_s\}$ $(s > r)$, such that $S_s = \phi$ for s sufficiently large, while G_S is a function on $\prod_{s\leq r} \Omega_s$ and $W_S' = \bigotimes_{s>r} w_{S_s}$ on $\prod_{s>r} G_{N_s}'$.

Thus

$$T(G) = \Sigma\, U(G_S) \otimes V(W_S') \text{ and } \|T(G)\|_1 \leq \|U\|\, \|\Sigma\, G_S \otimes V(W_S')\|_1$$

Also, by definition of the operators T_s, we get

$$V(W_S') = \prod_{s>r} \left(\tfrac{1}{s}\right)^{|S_s|} W_S',$$

Consequently

$$\|\Sigma\, G_S \otimes V(W_S')\|_2^2 = \sum_S \prod_{s>r} (1-2^{-s})\left(\tfrac{1}{s}\right)^{2|S_s|} \|G_S\|_2^2$$

$$\leq \frac{\|G\|_2^2}{(r+1)^2} + \|\mathcal{E}_r[G]\|_2^2$$

So, combining previous inequalities, we see that

$$\|T(G)\|_1 \leq \|T\| \frac{\|F\|_\infty}{r+1} + \|T\| \frac{\|F\|_\infty}{2^{r/2}} \leq \frac{2}{r} \|T\|\, \|F\|_\infty$$

which leads to the required estimation of $\|T(F)\|_1$.

<u>Proof of Th. 5.15</u> : Let (F_k) be weakly null in $L^1(\mu)$ and assume moreover $\|F_k - T(F_k)\|_1 \to 0$. By truncation and application of 5.16, we see that $\lim F_k = 0$ in norm.

So T fulfils (c) of 5.4. The $\mathcal{L}^1$-space B obtained from 5.4 will fail the RNP and have Schur property.

It may be interesting to notice here that apparently it is not clear to modify our example in order to obtain a nowhere convergent <u>diadic</u> tree. The non-RNP subspaces of L^1 constructed in [31] failed explicitely the diadic-tree-property.

APPENDIX 1

THE THREE SPACE QUESTION FOR $\mathcal{L}^1$ AND $\mathcal{L}^\infty$ SPACES

Assume X a Banach space and Y a subspace of X. In the spirit of 5.1, one has the property

$$X\ \mathcal{L}^1,\ X/Y\ \mathcal{L}^1 \Rightarrow Y\ \mathcal{L}^1$$

and the dual reformulation

$$X\ \mathcal{L}^\infty,\ Y\ \mathcal{L}^\infty \Rightarrow X/Y\ \mathcal{L}^\infty$$

Our purpose is to show that these implications are wrong if Y and X/Y are replaced.
From [85], we know that the problem is equivalent to the so called "ℓ^1-complementation question" and which can be stated as follows

PROBLEM 1 : Let μ and ν be measures and $T : L^1(\mu) \to L^1(\nu)$ an isomorphic embedding. Does there always exist a projection of $L^1(\nu)$ onto the range of T ?

It was stated in this form in [47], [49], [56], [85] and [87].
Let us briefly indicate one side of the relation. Take $X = L^1(\mu)$ and Y = range of T. If X/Y is an infinite dimensional $\mathcal{L}^1$-space, then Y = $(X/Y)^*$ is injective and therefore complemented in X^* by a projection P.
Defining $Q = I - P$, it is easily seen that Q^* gives a projection of X^{**} onto Y^{**}. Since Y is an $L^1(\mu)$-space, it will be complemented in its bidual and we conclude that Y is complemented in X.

Problem 1 has the following finite dimensional reformulation (cfr. [47]).

PROBLEM 2 : Does there exist for each $\lambda < \infty$ some $C < \infty$ such that given a finite dimensional subspace E of $L^1(\nu)$ satisfying $d(E,\ell^1(\dim E)) \leq \lambda$ (d = Banach-Mazur distance), one can find a projection $P : L^1(\nu) \to E$ with $\|P\| \leq C$?

In [47], L. Dor obtained a positive solution to problem 1 provided $\|T\|\ \|T^{-1}\| < \sqrt{2}$. It was shown by L. Dor and T. Starbird

(cfr. [49]) that any ℓ^1- subspace of $L^1(\nu)$ which is generated by a sequence of probabilistically independent random variables is complemented. A slight improvement of this result will be given in the remarks below, where we show that problem 2 is affirmative under the additional hypothesis that E is spanned by independent variables.
Our main purpose is to show that the general solution to the above questions is negative. Examples of uncomplemented ℓ^p-subspaces of L^p $(1 < p < \infty)$ were already discovered (see [109] for the cases $2 < p < \infty$ and $1 < p < 4/3$ and [8] for $1 < p < 2$).

In order to present the example, we use the notations introduced in section 6 of V. Thus G_N is the finite Cantor group with 2^N points, w_S denotes the S-Walsh-function and T_ε the convolution operator.
Before describing the example, we give some Lemma's.

<u>LEMMA 1</u> : If $f \in L^1(G_N)$, then $\|T_\varepsilon f\|_2 \leq |\int f \, dm_N| + \varepsilon \|f\|_2$

<u>Proof</u> : Take $f = a_\phi + \Sigma_{S \neq \phi} a_S w_S$ the Walsh expansion of f. Then

$$T_\varepsilon f = a_\phi + \Sigma_{S \neq \phi} a_S \varepsilon^{|S|} w_S$$

and hence $\|T_\varepsilon f\|_2^2 = |a_\phi|^2 + \Sigma_{S \neq \phi} |a_S|^2 \varepsilon^{2|S|} \leq |a_\phi|^2 + \varepsilon^2 \|f\|_2^2$.

The required inequality follows.

<u>LEMMA 2</u> : Let $f_1, \ldots, f_d$ be functions in $L^1(G_N)$ such that for each $i = 1, \ldots, d$

1. $\int f_i \, d\, m_N = 0$

2. $\int_{A_i} |f_i| \, d\, m_N \geq \delta \|f_i\|_1$ where $A_i = [|f_i| \geq d \|f_i\|_1]$

Then

$$\int_{G_N \times \ldots \times G_N} |f_1(x_1) + \ldots + f_d(x_d)| \, dm_N(x_1) \ldots dm_N(x_d) \geq \frac{\delta}{6} \Sigma_{i=1}^d \|f_i\|_1$$

Proof : For $i = 1,\dots,d$, take $B_i = G_N \setminus A_i$ and let C_i be the subset of $G_N \times \dots \times G_N$ defined by $C_i = B_1 \times \dots B_{i-1} \times A_i \times B_{i+1} \times \dots \times B_d$. Remark that $m_N(A_i) \leq \frac{1}{d}$ and hence $m_N(B_i) \geq 1 - \frac{1}{d}$. Let $\varepsilon_1,\dots,\varepsilon_d$ be Rademacher functions on $[0,1]$. By unconditionality, we get

$$\int_{G_N \times \dots \times G_N} |\Sigma_{i=1}^d f_i(x_i)| \, dm_N(x_1) \dots dm_N(x_d) \geq$$

$$\frac{1}{2} \int_0^1 \int_{G_N \times \dots \times G_N} |\Sigma_{i=1}^d \varepsilon_i(\omega) f_i(x_i)| \, dm_N(x_1) \dots dm_N(x_d) \, d\omega \geq$$

$$\frac{1}{2} \Sigma_i \int_{C_i} |f_i(x_i)| \, dm_N(x_1) \dots dm_N(x_d) \geq$$

$$\frac{1}{2} (1-\tfrac{1}{d})^{d-1} \Sigma_i \int_{A_i} |f_i(x) \, dm_N(x) \geq \frac{\delta}{6} \Sigma_i \|f_i\|_1, \text{ as required.}$$

We use the symbol $\oplus$ to denote the direct sum in ℓ^1-sense. For fixed N and d, take

$$X = \underbrace{L^1(G_N) \oplus \dots \oplus L^1(G_N)}_{d \text{ copies}} \quad \text{and} \quad Y = L^1(\underbrace{G_N \times \dots \times G_N}_{d \text{ factors}})$$

Consider the maps

$\alpha : X \to \ell^1(d)$

$\beta : X \to Y$

and for $0 \leq \varepsilon \leq 1$

$\gamma_\varepsilon : X \to X$

defined by

$$\alpha(f_1 \oplus \dots \oplus f_d) = (\int f_1 \, dm_N, \dots, \int f_d \, dm_N)$$

$$\beta(f_1 \oplus \dots \oplus f_d) = \Sigma_{i=1}^d (f_i(x_i) - \int f_i \, dm_N)$$

where $(x_1,\dots,x_d) \in G_N \times \dots \times G_N$ is the product variable

$$\gamma_\varepsilon(f_1 \oplus \dots \oplus f_d) = (f_1 - T_\varepsilon f_1) \oplus \dots \oplus (f_d - T_\varepsilon f_d)$$

Obviously $\|\alpha\| \leq 1$, $\|\beta\| \leq 2$ and $\|\gamma_\varepsilon\| \leq 2$.

Let $\Lambda_\varepsilon : X \to \ell^1(d) \oplus Y \oplus X$ be the map $\alpha \oplus \beta \oplus \gamma_\varepsilon$, clearly satisfying $\|\Lambda_\varepsilon\| \leq 5$.

LEMMA 3 : Under the above notations, $\|\Lambda_\varepsilon(\varphi)\| \geq \frac{1}{24} \|\varphi\|_1$ for each $\varphi \in X$, whenever $0 < \varepsilon \leq \frac{1}{4d}$.

Proof : Assume $\varphi = f_1 \oplus \ldots \oplus f_d$ and take for each $i = 1,\ldots,d$

$$g_i = f_i - \int f_i \, dm_N$$

$$A_i = [|g_i| \geq d\|g_i\|_1], \quad B_i = G_N \setminus A_i, \quad g_i' = g_i \chi_{A_i} \text{ and } g_i'' = g_i \chi_{B_i}$$

Let further $I = \{i = 1,\ldots,d \; ; \; \|g_i'\|_1 > \frac{1}{4}\|g_i\|_1\}$ and $J = \{1,\ldots,d\} \setminus I$.

Using lemma 2, we find that

$$\|\beta(f_1 \oplus \ldots \oplus f_d)\|_1 \geq \int_{G_N \times \ldots \times G_N} |\Sigma_{i\in I} \, g_i(x_i)| \, dm_N(x_1) \ldots dm_N(x_d) \geq$$

$$\frac{1}{24} \Sigma_{i\in I} \|g_i\|_1.$$

On the other hand, by lemma 1

$$\|T_\varepsilon g_i\|_1 \leq \|T_\varepsilon g_i'\|_1 + \left|\int g_i'' \, dm_N\right| + \varepsilon \|g_i''\|_2 \leq 2\|g_i'\|_1 + \varepsilon d \|g_i\|_1$$

and hence for $i \in J$

$$\|f_i - T_\varepsilon f_i\|_1 = \|g_i - T_\varepsilon g_i\|_1 \geq \|g_i\|_1 - \|T_\varepsilon g_i\|_1 \geq \tfrac{1}{4}\|g_i\|_1$$

Consequently $\|\gamma_\varepsilon(f_1 \oplus \ldots \oplus f_d)\|_1 \geq \Sigma_{i\in J} \|f_i - T_\varepsilon f_i\|_1 \geq \frac{1}{4} \Sigma_{i\in J} \|g_i\|_1$.

Combination of these inequalities leads to

$$\|\Lambda_\varepsilon(\varphi)\|_1 \geq \Sigma_{i=1}^d \left|\int f_i \, dm_N\right| + \frac{1}{24} \Sigma_{i=1}^d \|g_i\|_1 \geq \frac{1}{24} \Sigma_{i=1}^d \|f_i\|_1 =$$

$\frac{1}{24} \|\varphi\|_1$ proving the lemma.

COROLLARY 4 : Again under the above notations, denote R_ε the range of Λ_ε. Then $d(R_\varepsilon, \ell^1(d.2^N)) \leq \frac{1}{120}$ provided $0 < \varepsilon \leq \frac{1}{4d}$.

Our next aim is to show that R_ε is a badly complemented subspace of $\ell^1(d) \oplus Y \oplus X$ for a suitable choice of N, d and ε.

For each $\nu \in G_N$, we consider again the functions

$$\varepsilon_\nu = \prod_{n=1}^{N} (1 + \nu_n r_n)$$

on G_N, which form a system isometrically equivalent to the $\ell^1(2^N)$-basis and generating $L^1(G_N)$.
For a positive integer d we define also for $\nu \in G_N$

$$\xi_\nu = \frac{1}{d} \sum_{j=0}^{d-1} T_{\varepsilon^j}(e_\nu)$$

<u>LEMMA 5</u> : Fix any positive integer $d \geq 4$, take $N = d^{6d}$ and let $\varepsilon = \frac{1}{4d}$. Then $\|P\| \geq \frac{d}{384}$ for any projection P from $\ell^1(d) \oplus Y \oplus X$ onto R.

<u>Proof</u> : Take the functions ξ_ν as defined above.
If we take $A_\nu = [\xi_\nu > \frac{1}{4}]$, we get easily

$$m_N(A_\nu) \leq 2\, d\, (1 - \frac{\varepsilon^{2d}}{4})^{N/2} < \frac{1}{2}$$

by the choice of N and ε.

It follows that if $\psi_\nu = \xi_\nu - 1$, then

$$\|\psi\|_1 \geq \int_{A_\nu} \xi_\nu \, dm_N - m_N(A_\nu) \geq \int \xi_\nu \, dm_N - \frac{1}{4} - m_N(A_\nu) > \frac{1}{4}.$$

Assuming P a projection from $\ell^1(d) \oplus Y \oplus X$ onto R_ε, one may consider the operator $Q = \Lambda_\varepsilon^{-1} P$ from $\ell^1(d) \oplus Y \oplus X$ into X.

For each $i = 1,\ldots,d$ and $\nu \in G_N$, let φ_ν^i be ψ_ν seen as element of the ith component $L^1(G_N)$ in the direct sum X. Thus $\alpha(\varphi_\nu^i) = 0$, $\beta(\varphi_\nu^i) = \psi_\nu(x_i)$ and $\gamma(\varphi_\nu^i) = \varphi_\nu^i - T_\varepsilon(\varphi_\nu^i)$.

By well-known results concerning operators on L^1-spaces, we get

$$d \int \Sigma_\nu \; |\psi_\nu| \; dm_N =$$

$$\int \max_i \; (\Sigma_\nu \; |Q \; \Lambda_\varepsilon(\varphi_\nu^i)|) \; dm_N \oplus \ldots \oplus dm_N \leq$$

$$\int \max_i \; |Q| \; (\Sigma_\nu \; |\Lambda_\varepsilon(\varphi_\nu^i)|) \; dm_N \oplus \ldots \oplus dm_N \leq$$

$$\|Q\| \; \{\int \max_i \; (\Sigma_\nu |\psi_\nu(x_i)|) \; dm_N(x_1) \ldots dm_N(x_d) + \Sigma_i \Sigma_\nu \int |\varphi_\nu^i - T_\varepsilon(\varphi_\nu^i)| dm_N\}$$

Remark that, by symmetry, $\Sigma_\nu \; |\psi_\nu|$ is a constant function.
Because $\frac{1}{4} < \|\psi_\nu\|_1 \leq 2$ and $\|\psi_\nu - T_\varepsilon(\psi_\nu)\|_1 = \|\xi_\nu - T_\varepsilon(\xi_\nu)\|_1 =$

$\frac{1}{d} \|e_\nu - T_{\varepsilon^d}(e_\nu)\|_1 \leq \frac{2}{d}$, we find using lemma 3

$$d \; \Sigma_\nu \|\psi_\nu\|_1 \leq 24 \; \|P\| (\Sigma_\nu \; \|\psi_\nu\|_1 + 2^{N+1})$$

and hence

$$\|P\| \geq d \; \frac{\frac{1}{4} \; 2^N}{24(2^{N+1}+2^{N+1})} = \frac{d}{384}$$

completing the proof.

From corollary 4 and lemma 5, it follows that

THEOREM 6 : There exists a constant $0 < C < \infty$ such that whenever $\tau > 0$ and D is a positive integer which is large enough, one can find a D-dimensional subspace E of L^1 satisfying $d(E, \ell^1(D)) \leq C$ and $\|P\| \geq C^{-1} \; (\log \log D)^{1-\tau}$ whenever P is a projection from L^1 onto E.

This provides in particular a negative solution to problem 1 and problem 2 stated above.

This counterexample to the ℓ^1-complementation question is closely related to certain considerations about integrability moduli in L^1.

Following L. Dor, one may define local and uniform moduli for functions and subspaces of an $L^1(\mu)$-space.
For a function f in $L^1(\mu)$ and $\rho > 0$, take

$$\alpha(f,\rho) = \inf \{\mu(A) \; ; \int_A |f| \, d\mu \geq \rho \, \|f\|_1\}$$

If now E is a subspace of $L^1(\mu)$ and $\rho > 0$, let

$$\alpha(E,\rho) = \sup \{\alpha(f,\rho) \; ; \; f \in E\}$$

and

$$\beta(E,\rho) = \inf \{\mu(A) \; ; \int_A |f| \, d\mu \geq \rho \, \|f\|_1 \text{ for each } f \in E\}$$

Call $\alpha(E,\rho)$ a local modulus and $\beta(E,\rho)$ a uniform modulus of the space E.
Based on the ideas presented in the preceding section, the following can be proved.

LEMMA 7 : There exists a sequence (E_n) of finite dimensional subspaces of L^1 and constants $C < \infty$ and $c > 0$, such that

1. $d(E_n, \ell^1(\dim E_n)) \leq C$
2. $\lim_{n\to\infty} \alpha(E_n,c) = 0$
3. For each $\rho > 0$, $\inf_n \beta(E_n,\rho) > 0$.

As was pointed out by Dor [48], this leads to the existence of a non-complemented ℓ^1-subspace of L^1.

In fact, one may choose the spaces E_n of lemma 7 in such a way that they are well-complemented and probabilistically independent. This allows us to construct a non-complemented ℓ^1-direct sum of uniformly complemented, independent, uniform ℓ^1-isomorphs.
Thus the next result concerning independent functions can not be extended to independent ℓ^1-copies.

THEOREM 8 : If E is an ℓ^1-subspace of $L^1(\mu)$ spanned by independent variables, then E is complemented in $L^1(\mu)$ by a projection P whose norm $\|P\|$ can be bounded in function of $d(E,\ell^1(\dim E))$. (cfr. [49]).

There is an easy reduction to the case where E is generated by a sequence (f_k) of normalized, independent and mean zero variables. Using then the uniqueness up to equivalence of unconditional bases in ℓ^1-spaces (see [87]), it turns out that this sequence (f_k) is a "good" ℓ^1-bases for E, or more precisely there is some constant $M < \infty$, M only depending on $d(E,\ell^1(\dim E))$, so that

$$M^{-1} \Sigma_k |a_k| \leq \|\Sigma_k a_k f_k\| \leq \Sigma_k |a_k|$$

whenever (a_k) is a finite sequence of scalars.

Assume $\mathcal{E}_k$ (k=1,2,...) independent σ-algebra's such that f_k is $\mathcal{E}_k$-measurable. The main ingredient of the next lemma is the result [49].

LEMMA 9 : There exists a sequence of μ-measurable sets satisfying

1. $A_k \in \mathcal{E}_k$ for each k
2. $\int_{A_k} f_k \, d\mu \geq \rho$ for each k
3. $\Sigma_k \mu(A_k) \leq K$

where $\rho > 0$ and $K < \infty$ only depend on M and hence only on $d(E,\ell^1(\dim E))$.

The proof of this lemma is contained in [49], section 3. So we will not give it here. Let us now pass to the

Proof of Theorem 8 : We may clearly make the additional assumption that $\mu(A_k) < \frac{1}{3}$.

For each k, let $\mathcal{F}_k = G(\mathcal{E}_1,\ldots,\mathcal{E}_k)$ the σ-algebra generated by $\mathcal{E}_1,\ldots,\mathcal{E}_k$.
Take

$$B_1 = A_1 \quad \text{and} \quad B_k = A_k \setminus \cup_{\ell<k} A_\ell \quad \text{for } k > 1.$$

Clearly $B_k \in F_k$ for each k. Remark also that

$$\int_{B_k} f_k \, d\mu = \int f_k \chi_{A_k} \Pi_{\ell<k} (1-\chi_{A_\ell}) = \Pi_{\ell<k} (1-\mu(A_\ell)) \int_{A_k} f_k$$

and hence

$$\int_{B_k} f_k \, d\mu = c_k \geq \exp(-3K)\rho.$$

Define

$$\Delta_1[f] = E[f|F_1] \text{ and } \Delta_k[f] = E[f|F_k] - E[f|F_{k-1}] \text{ for } k > 1.$$

Thus

$$\Delta_k[f_\ell] = \delta_{k,\ell} f_\ell.$$

Next, take $P : L^1(\mu) \to E$ given by $P(f) = \Sigma_k c_k^{-1} < \Delta_k[f], B_k > f_k$

It is clear that P is a projection. We estimate its norm

$$\|P\| \leq \|\Sigma_k c_k^{-1} |\Delta_k [\chi_{B_k}]|\|_\infty$$

$$\leq \frac{\exp 3K}{\rho} \|\Sigma_k \chi_{B_k} + \Sigma_k \mu(A_k)\|_\infty$$

$$\leq (1+K) \frac{\exp 3K}{\rho}$$

Our example leaves the following questions unanswered

PROBLEM 3 : What is the biggest λ such that problem 1 has a positive solution provided $\|T\| \|T^{-1}\| < \lambda$?

For E subspace of L^1, define

$$\pi(E) = \inf \{\|P\| \; ; \; P : L^1 \to E \text{ is a projection}\}$$

Take further for fixed $n = 1,2,\ldots$ and $\lambda < \infty$

$$\gamma(n,\lambda) = \sup \{\pi(E) \; , \; \dim E = n \text{ and } d(E,\ell^1(n)) \leq \lambda\}$$

PROBLEM 4 : Find estimations on the numbers $\gamma(n,\lambda)$. At this point, it does not seem even clear that for fixed $\lambda < \infty$ the following holds $\lim_{n\to\infty} \frac{\gamma(n,\lambda)}{\sqrt{n}} = 0$.

Problem 1 seems so far unsolved in the translation invariant setting. Thus

PROBLEM 5 : Assume G a compact Abelean group and E a translation invariant subspace of $L^1(G)$. Is the orthogonal projection on E L^1-bounded if we assume E isomorphic to $L^1(G)$?

Related to this question is the following one, due to G. Pisier [102].

PROBLEM 6 : Let G be the Cantor group and define E as the subspace of $L^1(G)$ generated by the Walsh-functions w_S where $|S| \geq 2$. Obviously, E is uncomplemented. What about the following

a. Is E a $\mathcal{L}^1$-space ?

b. Is E isomorphic to $L^1(G)$?

Easy modifications of the construction given in the second section also allow us to obtain badly complemented $\ell^p(n)$-subspaces of L^p for $1 < p < 2$.

Although ℓ^1-sequences in L^1 are uncomplemented in certain cases, one can always decompose the sequence in a bounded number of complemented subsequences. This fact is basically a consequence of following elementary combinatorial principle.

LEMMA 9 : Assume (a_{ij}) an infinite matrix with positive elements and $C < \infty$ a constant such that $\Sigma_j a_{ij} < C$ for each i. Then, for given $\delta > 0$, one can partition the positive integers in finitely many subsets D fulfilling the condition

$$\sum_{j\in D, j\neq i} a_{ij} \leq \delta \quad \text{for each } i \in D.$$

Moreover, the number of sets D only depends on C and δ.

Proof : We first deal with the case of a finite matrix $(a_{ij})_{1\leq i,j\leq N}$. We decompose $\{1,2,\ldots,N\}$ in a bounded number of subsets D so that

$$\sum_{j\in D, j<i} a_{ij} < \delta/2 \text{ for } i \in D \qquad (*)$$

Doing this for the matrix (b_{ij}), where $b_{ij} = a_{N-i,N-j}$, provides a decomposition in sets E satisfying

$$\sum_{j\in E, j>i} a_{ij} < \delta/2 \text{ for } i \in E \qquad (**)$$

The required partition is then obtained by considering all sets of the form $D \cap E$.
Let us now show how the partition in sets satisfying (*) is obtained. Define D_1 by taking $1 \in D_1$ and, once $D_1 \cap \{1,2,\ldots,i-1\}$ determined, $i \in D_1$ iff

$$\sum_{j\in D_1, j<i} a_{ij} < \delta/2$$

This implies in particular $\sum_{j\in D_1} a_{ij} \geq \delta/2$ for all $i \notin D_1$.

We start again with the increasing enumeration of $\{1,2,\ldots,N\}\setminus D_1$ and find a set D_2 satisfying (*) and $\sum_{j\in D_2} a_{ij} \geq \delta/2$ if $i \notin D_1 \cup D_2$.

Iterating the procedure gives a partition

$$\{1,2,\ldots,N\} = D_1 \cup D_2 \cup \ldots \cup D_r$$

in sets satisfying (*).
The bound on r follows from the fact that for $i \in D_r$

$$\sum_{j\in D_s} a_{ij} \geq \delta/2 \quad \text{for} \quad s = 1,\ldots,r-1$$

and hence

$$C \geq \Sigma_j \, a_{ij} \geq \Sigma_{s=1}^{r-1} \Sigma_{D_s} \, a_{ij} \geq (r-1)\delta/2.$$

It remains to deal with the case of an infinite matrix (a_{ij}).
For each N, the (NxN)-matrix $(a^N_{ij}) = (a_{ij})_{1\leq i,j\leq N}$ allows a partition function

$$\tau_N : \{1,\ldots,N\} \to \{1,\ldots,t\}$$

such that

$$\Sigma_j \{a^N_{ij} \ ; \ j \neq i \text{ and } \tau_N(j) = \tau_N(i)\} \leq \delta \text{ for all } 1 \leq i \leq N$$

(the integer t depends only on C and δ and thus not on N).
Let $\mathcal{U}$ be a free ultra-filter on $\mathbb{N}$ and $\tau : \mathbb{N} \to \{1,\dots,t\}$ the function $\lim_{\mathcal{U}} \tau_N$. The partition of $\mathbb{N}$ induced by τ clearly satisfies the condition of the lemma.

<u>LEMMA 10</u> : Let (x_n) and (x_n^*) be sequences in X and X^* for which there exist constants $M < \infty$ and $0 \leq \lambda < 1$ such that

(i) $\|x_n\| \leq M$

(ii) $\|\Sigma' \varepsilon_n x_n^*\| \leq M$ $(\varepsilon_n = 1,-1,0)$

(iii) $\langle x_n, x_n^* \rangle = 1$

(iv) $\sum_{n\neq m} |\langle x_m, x_n^* \rangle| \leq \lambda$ for each m.

Then (x_n) is a $(1-\lambda)/_{2M}$ ℓ^1-sequence and $[x_n]$ is complemented in X by a projection P with $\|P\| \leq 2M^2/_{1-\lambda}$.

<u>Proof</u> : Denote for convenience $E = [x_n]$. Define a linear map

$$U : X \to \ell^1 \text{ by } U(x) = (\langle x, x_n^* \rangle)$$

From (ii), $\|U\| \leq 2M$. Also, if we let $\xi_n = U(x_n)$, by (iii) and (iv)

$$\|\Sigma a_m \xi_m\|_1 = \Sigma_n |\langle \Sigma a_m \xi_m, x_n^* \rangle|$$

$$\geq \Sigma_n (|a_n| \, |\langle x_n, x_n^* \rangle| - \sum_{m\neq n} |a_m| \, |\langle x_m, x_n^* \rangle|)$$

$$\geq \Sigma_n |a_n| - \Sigma_m |a_m| \, (\sum_{n\neq m} |\langle x_m, x_n^* \rangle|) \geq (1-\lambda) \, \Sigma |a_n|$$

Thus we may consider the map

$$V : [\xi_n] \to \ell^1, \ V(\xi_n) = e_n$$

where e_n stands for the nth unit vector.
Define

$$W : \ell^1 \to E \text{ by } W(e_n) = x_n$$

Since

$$\| \Sigma\, a_m\, x_m \| \geq \frac{1 - \lambda}{2M} \Sigma\, |a_m|$$

and P = W V U is a projection from X onto E, we get the required conclusion.

PROPOSITION 11 : A sequence (x_n) in a complex Banach space X is a finite union of complemented ℓ^1-sequences if and only if there is a sequence (x_n^*) in X^* and $M < \infty$ for which (i), (ii), (iii) of lemma 10 hold.

Proof of Prop. 11 : Fix $\lambda < 1$. By lemma 10, we only have to partition $\mathbb{N}$ in finitely many sets D fulfilling the condition

$$\sup_{m \in D} \sum_{n \in D, n \neq m} |\langle x_m, x_n^* \rangle| \leq \lambda$$

This is done by application of lemma 2 to the matrix $(a_{m,n})$ where $a_{m,n} = |\langle x_m, x_n^* \rangle|$. It is indeed clear that for each m

$$\Sigma_n' |\langle x_m, x_n^* \rangle| \leq \|x_m\|\, \|\Sigma\, c_n\, x_n^*\| \leq 2\|x_m\|\, \|\Sigma\, \varepsilon_n\, x_n^*\| \leq 2M^2,$$

for a suitable choice of complex numbers c_n with $|c_n| = 1$ and $\varepsilon_n = 1,-1$.

We apply Prop. 11 in the case X is an $L^1(\mu)$-space. First, let us recall following basic lemma, for which we refer again to [47].

LEMMA 12 : If (f_n) is an ℓ^1-sequence in $L^1(\mu)$, then there exist disjoint μ-measurable sets A_n such that $\int_{A_n} |f_n|\, d\mu \geq \delta$ for each n, for some $\delta > 0$.

Combining Prop. 11 and Lemma 12 gives clearly

PROPOSITION 13 : A bounded sequence (f_n) in $L^1(\mu)$ is the finite union of complemented ℓ^1-sequences provided the f_n have at least mass $\delta > 0$ on disjoint μ-measurable sets.

Hence

COROLLARY 14 : Any ℓ^1-sequence in an $L^1(\mu)$-space is a finite union of ℓ^1-sequences with complemented linear span.

Our next purpose is to sketch briefly an application of Prop. 11 in interpolation theory.

Let Δ be a set and $\ell^\infty(\Delta)$ the algebra of complex bounded functions on Δ equipped with the sup norm. Say that a subalgebra $\mathfrak{A}$ of $\ell^\infty(\Delta)$ is C-interpolating for a sequence (z_m) in Δ provided for any bounded sequence (a_m) of complex numbers, there exists $f \in \mathfrak{A}$ satisfying

(i) $f(z_m) = a_m$ for each m

(ii) $\|f\| \leqslant C\|(a_m)\|_\infty$.

The next lemma (cfr. [27]) is a generalized version of a result of P. Beurling (see [51]) in the case $\Delta = \{z \in \mathbb{C}\ ;\ |z| < 1\}$ and $\mathfrak{A} = H^\infty$.

LEMMA 15 : If (z_m) is a finite C-interpolating sequence, then there are elements φ_m in $\mathfrak{A}$ such that

(i) $\varphi_m(z_n) = \delta_{m,n}$ for all m,n

(ii) $\|\Sigma\ |\varphi_m|\| \leqslant C'$

where the constant C' only depends on C.

We include a simple proof, which was communicated to the author by E. Amar.

Proof of Lemma 15 : Let $z_0, z_1, \ldots, z_{N-1}$ be the points and denote $\mathbb{Z}_N$ the cyclic group of Nth unit roots $\nu_0, \nu_1, \ldots, \nu_{N-1}$. By hypothesis, one can find for each m some $f_m \in \mathfrak{A}$ satisfying the conditions

$$\|f_m\|_\infty \leqslant C \quad \text{and} \quad f_m(z_n) = (\nu_m)^n$$

Define

$$\alpha_m = \frac{1}{N} \Sigma_k (\bar{\nu}_k)^m f_k \quad \text{and} \quad \varphi_m = \alpha_m^2$$

which are members of $\mathfrak{A}$. First notice that

$\alpha_m(z_n) = \frac{1}{N} \Sigma_k (\bar{\nu}_k)^m (\nu_k)^n = \delta_{m,n}$ and hence also $f_m(z_n) = \delta_{m,n}$.

Next, if we fix $z \in \Delta$ and see $(f_m(z))_{0 \leq m < N}$ as a function on $\mathbb{Z}_N$, application of Plancherel gives

$$\Sigma_m |\alpha_m(z)|^2 = \Sigma_m |\frac{1}{N} \Sigma_k (\bar{\nu}_k)^m f_k(z)|^2 \leq C^2,$$

which completes the proof.

Denote for $z \in \Delta$ by $\tilde{z} \in \ell^\infty(\Delta)^*$ the evaluation functional. Lemma 15 means that if $(\tilde{z}_m)$ is a finite ℓ^1-sequence in $\mathfrak{A}^*$, then $[\tilde{z}_m]$ is uniformly complemented in $\mathfrak{A}^*$.

Based on lemma 15, it is easily seen that a union of two interpolating sequences which are uniformly separated in $\mathfrak{A}^*$, remains interpolating. More precisely

LEMMA 16 : Assume (z_m) and (w_n) finite C-interpolating sequences so that $\inf_{m,n} \|\tilde{z}_m - \tilde{w}_m\|_{\mathfrak{A}^*} > \delta > 0$. Then $(z_m) \cup (w_n)$ is a C'-interpolating sequence with C' only dependent on C and δ.

Proof : Assume (φ_m) and (ψ_n) sequences in $\mathfrak{A}$ satisfying Lemma 15 with respect to the respective points (z_m) and (w_n). For each pair (m,n), there is also a function $\alpha_{m,n} \in \mathfrak{A}$ for which $\alpha_{m,n}(z_m) = 0$ and $\alpha_{m,n}(w_n) = \delta$ ($\|\alpha_{m,n}\| \leq 1$). If now (a_m) is a sequence of complex numbers, then the formula

$$f = \Sigma_m a_m \varphi_m \{1 - \Sigma_n \delta^{-1} \alpha_{m,n} \psi_n\}$$

yields a function in $\mathfrak{A}$ taking value a_m in z_m and which is zero on $\{w_n\}$. The verification of the uniform bound is immediate.

Combining Proposition 11 and Lemma 16, the following is obtained

PROPOSITION 17 : Assume (z_m) a finite sequence in Δ, (φ_m) a sequence in $\mathfrak{A}$, $M < \infty$ and $\delta > 0$ such that

(i) $\inf_{m \neq n} \|\tilde{z}_m - \tilde{z}_n\|_{\mathfrak{A}^*} > \delta$

(ii) $\|\Sigma |\varphi_m|\| \leqslant C$

(iii) $\varphi_m(z_m) = 1$

Then (z_m) is C-interpolating, C only dependent on M and δ.

We apply prop. 17 with $X = \mathfrak{A}^*$, $x_m = \tilde{z}_m$ and $x_m^* = \varphi_m$.

Of course, the last three results generalize to infinite sequences as soon as $\mathfrak{A}$ is w^*-closed in $\ell^\infty(\Delta)$.

This is in particular the case if we take for $\mathfrak{A}$ the algebra H^∞ on the unit disc. For more details on interpolating sequences in the disc, the reader is referred to [38] and [51].

Conditions (ii), (iii) of Prop. 17 are equivalent to the fact that the measure μ on Δ given by

$$\mu = \Sigma_m (1-|z_m|)\, \tilde{z}_m$$

is a so-called Carleson measure. Again, we refer to [51] for definitions and details. The distance induced by $\mathfrak{A}$ is given by

$$\|\tilde{z} - \tilde{w}\|_{\mathfrak{A}^*} = \left|\frac{z-w}{1-\bar{z}w}\right| \quad \text{for } z,w \in \Delta$$

Prop. 11 extends the fact that Carleson-sequences are finite unions of interpolating sequences and Prop. 17 says that uniformly separated Carleson sequences are interpolating (cfr. [117]).

In [27], the following general result is shown (cfr. also [26]).

THEOREM 18 : To each $\delta > 0$ corresponds $\delta_1 > 0$ such that if $f_1, f_2, \ldots, f_n$ are positive norm-1 functions on the circle T and satisfy

(i) $\int \max \lambda_m f_m \geq \delta \Sigma \lambda_m$ for all $\lambda_m \geq 0$,

then there are H^∞-functions $g_1, g_2, \ldots, g_n$ for which

(ii) $|g_1| + |g_2| + \ldots + |g_n| \leq 1$

(iii) $\langle f_m, g_m \rangle = \int f_m g_m = \delta_1$.

Weakening of the definition of interpolating sequence in the unit disc by requiring the interpolating function only to be a bounded harmonic function on Δ leads to the notion of harmonically interpolating sequence. This property for a sequence (z_m) in $\{z \in \mathbb{C} ; |z| < 1\}$ is equivalent to the fact that the sequence of Poisson kernels (P_{z_m}), $P_z(\theta) = \dfrac{1 - |z|^2}{|1-\bar{z}e^{i\theta}|^2}$ is equivalent to the usual ℓ^1-basis in the space $L^1(\Pi)$.
Thus, if (z_m) is harmonically interpolating, (P_{z_m}) fulfils (i) of Th. 18. This provides however H^∞-functions (g_m) for which

(ii') $|g_1| + |g_1| + \ldots + |g_n| \leq$ const.

(iii') $g_m(z_m) = \langle P_{z_m}, g_m \rangle = 1$.

Since now a theorem of Harnack asserts that for $z, w \in \Delta$

$$\|P_z - P_w\|_1 \sim \|\tilde{z} - \tilde{w}\|_{(H^\infty)^*}$$

Prop. 17 gives that (z_m) is actually interpolating. Hence

<u>THEOREM 19</u> : For sequences in the open unit disc, the notions "interpolating" and "harmonically interpolating" are equivalent.

This result is due to J. Garnett [59].

APPENDIX 2

DUNFORD-PETTIS PROPERTY OF L_C^1 AND RELATED SPACES

1. INTRODUCTION

If X is a Banach space, denote C_X the space of continuous X-valued functions and L^1_X the space of Bochner-integrable X-valued function (classes).

It is not yet known if the D-P property of X implies the D-P property of C_X and L^1_X. The results presented here are related to this problem. More precisely, we will show that C_{L^1} and L^1_C-spaces and and also their duals are D-P, which solves an open problem. As will be clear from what follows, the argument here is more involving than for C(K) and $L^1(\mu)$ spaces and uses a "non-linear" technique. Our results also allow us to show that l^2 is not finite dimensionally complemented in C_{L^1}, a problem raised in [57].

We will use the notation (ε_i) for the sequence of Rademacker functions on [0,1]. The next result is a particular case of a more general theorem (see [18] and [101]) and is in fact easily derived from the Rosenthal characterization theorem (see 104).

PROPOSITION 1 : If X is a Banach space and $(x_i)_{i=1,2,...}$ a WCC sequence in X, then the sequence $(x_i \otimes \varepsilon_i)_{i=1,2,...}$ is WCC in L^1_X.

The only interest of prop. 1 here is the simplification of some of the arguments.

2. THE BASIC RESULT

Let X be a Banach space. Denote $\oplus_\infty X$ the l^∞-sum of spaces X.
If $\xi \in \oplus_\infty X$, take $|\xi| = (\|\xi_1\|, \|\xi_2\|, \ldots)$ which is an element of l^∞. For $\eta \in [\oplus_\infty X]^*$, we let $|\eta|$ be the positive element of $(l^\infty)^*$, given by $\langle\alpha,|\eta|\rangle = \sup\{\langle\xi,\eta\rangle ; \xi \in \oplus_\infty X \text{ and } |\xi| \leq \alpha\}$ for $\alpha \in l^\infty_+$.
It is indeed clear that $|\eta|$ is linear and $\||\eta|\| = \|\eta\|$.

PROPOSITION 2 : If A is a WCC subset of $[\oplus_\infty X]^*$, then
$\{|\eta| ; \eta \in A\}$ is equi-continuous.

Proof : Suppose $\{|\eta| \; ; \; \eta \in A\}$ not equi-continuous. Then there exist $\varepsilon \geq 0$, a sequence (η_i) in A and a sequence (M_i) of subsets of $\mathbb{N}$, such that $|\eta_i|(M_i) > \varepsilon$ and $|\eta_i|(M_j) < 3^{-j}\varepsilon$ for $j > i$. We will obtain a contradiction on the WCC of A by showing that $(\eta_i \otimes \varepsilon_i)$ is an l^1-basis. So fix an integer k and choose scalars $a_1,\dots,a_k$. Define $N_i = M_i \setminus \cup_{j=i+1}^k M_j$ for $i = 1,\dots,k$. Clearly

$$|\eta_i|(N_i) \geq |\eta_i|(M_i) - \Sigma_{j=i+1}^k |\eta_i|(M_j) > \varepsilon - \Sigma_{j=i}^k 3^{-j} \varepsilon > \frac{\varepsilon}{2}.$$

Now

$$\int \|\Sigma_{i=1}^k a_i \, \varepsilon_i(\omega) \, \eta_i\| d\omega \geq \Sigma_{j=1}^k \int |\Sigma_{i=1}^k a_i \, \varepsilon_i(\omega) \, \eta_i|(N_j) \, d\omega$$

$$\geq \Sigma_{j=1}^k |a_j| \, |\eta_j|(N_j) \geq \frac{\varepsilon}{2} \Sigma_{j=1}^k |a_j|,$$ proving the lemma.

The Banach space X will be $L^1(\mu)$, where μ is some probability measure.

PROPOSITION 3 : Let A be a WCC subset of $\oplus_\infty L^1$, ν a positive element of $(l^\infty)^*$ and $\varepsilon > 0$. Then there exist some ζ in $\oplus_\infty L^1_+$, $\|\zeta\| = 1$ and some $\delta > 0$ such that $\nu(M) < \varepsilon$ for each $\xi \in A$, where $M = \{n \in \mathbb{N} \; ; \; \int_S \zeta_n \, d\mu < \delta$ and $\int_S \xi_n \, d\mu > \varepsilon$ for some $S\}$.

Proof : The requirement $\|\zeta\| = 1$ is of course unimportant. Assume the claim untrue. Proceeding by induction, it is possible to obtain a sequence (ξ^i) in A such that $\nu(M_i) \geq \varepsilon$ for each i, where $M_i = \{n \in \mathbb{N} \; ; \; \int_S \zeta_n^i \, d\mu < 3^{-i}\varepsilon$ and $\int_S |\xi_n^i| \, d\mu > \varepsilon$ for some $S\}$ and

$\zeta^i \in \oplus_\infty L^1$ is given by $\zeta_n^i = \max\,(|\xi_n^j| \; ; \; 1 \leq j \leq i-1)$.

The construction is again completely straightforward, so we omit details. There is an infinite subset I of $\mathbb{N}$ such that $(M_i)_{i \in I}$ has the finite intersection property. We will show that $(\xi^i \otimes \varepsilon_i)_{i \in I}$ is equivalent to the l^1-basis, which will provide the required contradiction.

Fix $i_1 < i_2 < \dots < i_r$ in I and scalars $a_1, a_2, \dots, a_r$. Take $n \in \cap_{s=1}^r M_{i_s}$. For each $s = 1,\dots,r$ there exist a set S_s satisfying

$$\int_{S_s} |\xi_n^{i_s}| \, d\mu > \varepsilon \text{ and } \int_{S_s} |\xi_n^i| \, d\mu < 3^{-s} \varepsilon \text{ for } i < i_s.$$

So if $T_s = S_s \setminus \cup_{t=s+1}^r S_t$, we obtain

$$\int_{T_s} |\xi_n^{i_s}|\, d\mu \geq \int_{S_s} |\xi_n^{i_s}|\, d\mu - \Sigma_{t=s+1}^{r} \int_{S_t} |\xi_n^{i_s}|\, d\mu >$$

$$\varepsilon - \Sigma_{t=s+1}^{r} 3^{-t}\, \varepsilon > \frac{\varepsilon}{2}.$$

Finally $\int \| \Sigma_s\, a_s\, \varepsilon_s(\omega)\, \xi^{i_s} \| d\omega \geq \int \| \Sigma_s\, a_s\, \varepsilon_s(\omega)\, \xi_n^{i_s} \| d\omega$

$$\geq \Sigma_{t=1}^{r} \int \int_{T_t} |\Sigma_s\, a_s\, \varepsilon_s(\omega)\, \xi_n^{i_s}|\, d\mu\, d\omega$$

$$\geq \Sigma_{t=1}^{r} |a_t| \int_{T_t} |\xi_n^{i_t}| d\mu \geq \frac{\varepsilon}{2}\, \Sigma_{t=1}^{r} |a_t|.$$

This completes the proof.

THEOREM 4 : $\oplus_\infty L^1$ has the D-P property.

We assume (ξ^i) a weakly null sequence in $\oplus_\infty L^1$, (η_i) a weakly null sequence in $(\oplus_\infty L^1)^*$ such that $\|\xi^i\| = \|\eta_i\| = 1$ and $\langle \xi^i, \eta_i \rangle > \rho$ $(\rho > 0)$ and will work towards a contradiction.

If $\mathcal{E}$ is a sub-σ-algebra, $\mu' \ll \mu$ and $f \in L^1(\mu')$, then $E[f, \mathcal{E}, \mu']$ will denote the conditional expectation of f with respect to $\mathcal{E}$ in the space $L^1(\mu')$.

LEMMA 5 : There exist θ in $\oplus_\infty L^1_+$, a sequence (f^j) in $\oplus_\infty L^\infty$ and a sequence (λ_j) in $(\oplus_\infty L^1)^*$ such that the following conditions are fulfilled

1. $\|f^j\| = \sup_n \|f^j_n\|_\infty \leq 1$
2. For each n, the sequence $(f^j_n)_j$ is a martingale difference sequence in the space $L^1(\mu_n)$, where $d\mu_n = \theta_n\, d\mu$
3. $\langle \Delta^j, \lambda_j \rangle > \frac{\rho}{4}$, where $\Delta^j \in \oplus_\infty L^1$ is defined by $\Delta^j_n = f^j_n\, \theta_n$
4. (λ_j) is a subsequence of (η_i)

Proof : Let $\nu_i = |\eta_i|$ and $\nu = \Sigma_i\, 2^{-i}\, \nu_i$. Take $\iota = \frac{\rho}{12}$.
By prop. 2, there is some $0 < \varepsilon < \iota$ so that $\sup_i \nu_i(M) < \iota$ if $M \subset \mathbb{N}$ and $\nu(M) < \varepsilon$. Let then ζ and δ be as in lemma 2, applied with $A = \{\xi^i\, ;\, i\}$.

Take $\theta = \delta^{-1}\zeta$. For each n and i, let $S_n^i = [\,|\xi_n^i| > \theta_n]$ and $\psi_n^i = (1-\chi_n^i)\xi_n^i$, where χ_n^i is the characteristic function of S_n^i.

Let also $M_i = \{n \; ; \|\xi_n^i - \psi_n^i\|_1 > \varepsilon\}$ for each i.

Clearly $\int_{S_n^i} \zeta_n \, d\mu < \delta \|\xi_n^i\|_1 \leq \delta$ and $\|\xi_n^i - \psi_n^i\|_1 = \int_{S_n^i} |\xi_n^i| \, d\mu$.

So $\nu(M_i) < \varepsilon$, by the choice of ζ and δ.

Since $\psi_n^i \leq \theta_n$, we may write $\psi_n^i = g_n^i \theta_n$ for some $g_n^i \in L^\infty(\mu)$; $\|g_n^i\|_\infty \leq 1$.

We construct the f^j and λ_j inductively.

Suppose f^j obtained such that each f_n^j is $\mathcal{E}_n^j$-measurable for some finite algebra $\mathcal{E}_n^j$, where the number $d = d_j$ of atoms of $\mathcal{E}_n^j$ does not depend on n. Denote $I_{n,e}$ $(1 \leq e \leq d)$ the atoms of $\mathcal{E}_n^j$ and consider for each $e = 1,\ldots,d$ the weakly null sequence $(\alpha^i(e))_i$ in l^∞, defined by $\alpha_n^i(e) = \int_{I_{n,e}} \xi_n^i \, d\mu$.

Let B be the direct sum of d copies of the space l^∞. We may introduce the operator $\Gamma : B \to \oplus_\infty L^1(\mu)$, defined by

$$\Gamma_n(\beta(1),\ldots,\beta(d)) = \Sigma_{e=1}^d \frac{\beta_n(e)}{\mu_n(I_{n,e})} \pi_{n,e} \, \theta_n \, ,$$

where $\pi_{n,e}$ is the characteristic function of $I_{n,e}$. It is indeed clear that $\|\Gamma_n(\beta(1),\ldots,\beta(d))\|_1 \leq \Sigma_e |\beta_n(e)| \leq \Sigma_e \|\beta(e)\|$ and hence Γ is bounded.

The sequence $(\Gamma^*(\eta_i))$ is weakly null in B^*. Since l^∞ is D-P, see [8], B is also D-P and therefore $\langle\alpha^i(1),\ldots,\alpha^i(d),\Gamma^*(\eta_i)\rangle$ tends to 0 for $i \to \infty$.

Fix some i large enough to ensure that $|\langle\sigma^i,\eta_i\rangle| < \varepsilon$, where $\sigma^i = \Gamma(\alpha^i(1),\ldots,\alpha^i(d))$. Remark that $\|\sigma_n^i\| \leq \Sigma_e |\alpha_n^i(e)| \leq \|\xi_n^i\|$ and thus $\|\sigma^i\| \leq 1$.

A standard argument allows us to construct finite algebra's $(\mathcal{E}_n^{j+1})$ such that $\mathcal{E}_n^j \subset \mathcal{E}_n^{j+1}$, $\|g_n^i - E[g_n^i, \mathcal{E}_n^{j+1}, \mu_n]\|_\infty \leq \frac{\varepsilon}{\|\mu_n\|}$ and the number of atoms of $\mathcal{E}_n^{j+1}$ is bounded.

Define f^{j+1} by $2f_n^{j+1} = E[g_n^i, \mathcal{E}_n^{j+1}, \mu_n] - E[g_n^i, \mathcal{E}_n^j, \mu_n]$ if $n \in N_i = \mathbb{N} \setminus M_i$ and $f_n^{j+1} = 0$ if $n \in M_i$.

Take $\lambda_{j+1} = \eta_i$.
It remains to verify (3).
For each n,

$$\sigma_n^i - E[g_n^i, \mathcal{E}_n^j, \mu_n]\theta_n =$$

$$\Sigma_e \frac{\int (\xi_n^i - g_n^i \theta_n)\pi_{n,e}\, d\mu}{\mu_n(I_{n,e})} \pi_{n,e}\, \theta_n$$

and therefore $\|\sigma_n^i - E[g_n^i, \mathcal{E}_n^j, \mu_n]\theta_n\|_1 \leqslant \|\xi_n^i - \psi_n^i\|_1$.

So $<2\Delta^{j+1}, \lambda_{j+1}> = <(2f_n^{j+1}\, \theta_n)_{n\in N_i}, \eta_i> \geqslant$

$<(g_n^i\, \theta_n)_{n\in N_i}, \eta_i> - <(\sigma_n^i)_{n\in N_i}, \eta_i>$

$- \sup_{n\in N_i} \|g_n^i\, \theta_n - E[g_n^i, \mathcal{E}_n^{j+1}, \mu_n]\theta_n\|_1$

$- \sup_{n\in N_i} \|\sigma_n^i - E[g_n^i, \mathcal{E}_n^j, \mu_n]\theta_n\|_1$

$\geqslant <(\psi_n^i)_{n\in N_i}, \eta_i> - <(\sigma_n^i)_{n\in N_i}, \eta_i> - 2\varepsilon$

$\geqslant <\xi^i, \eta_i> - <\sigma^i, \eta_i> - <(\xi_n^i)_{n\in M_i}, \eta_i> -$

$<(\sigma_n^i)_{n\in M_i}, \eta_i> - 3\varepsilon$

$\geqslant \rho - (\|\xi^i\| + \|\sigma^i\|)\nu_i(M_i) - 4\varepsilon \geqslant \rho - 2\iota - 4\varepsilon$

or $<\Delta^{j+1}, \lambda_{j+1}> \geqslant \frac{\rho}{4}$. This proves the lemma.

<u>Proof of theorem 4</u> : Take $0 < \kappa < \frac{\rho}{4\|\theta\|}$. Since (λ_j) is weakly null, it is possible to find a finitely supported sequence (a_j) of positive scalars such that $\Sigma\, a_j = 1$ and $\|\Sigma_j\, a_j\, \varepsilon_j\, \lambda_j\| \leqslant \kappa$ for all signs $\varepsilon_j = \pm 1$.
Consequently

$$\int \|\Sigma_j\, a_j\, \varepsilon_j(\omega)\lambda_j\| d\omega \leqslant \kappa$$

Take now for each ω the (Riesz) product

$$R_n(\omega) = \Pi_j(1 + \varepsilon_j(\omega) f_n^j)\theta_n$$

Clearly $R_n(\omega)$ is a positive function and $\|R_n(\omega)\|_1 = \int R_n(\omega)d\mu = \|\theta_n\|_1$, using the fact that

$$\int f_n^{j_1} f_n^{j_2} \dots f_n^{j_r} d\mu_n = 0$$

whenever $j_1 < j_2 < \dots < j_r$.

So $R(\omega) = (R_1(\omega), R_2(\omega), \dots)$ is a member of $\oplus_\infty L^1$ and $\|R(\omega)\| = \|\theta\|$. Therefore

$$\frac{\rho}{4} > \int \|\Sigma_j a_j \varepsilon_j(\omega)\lambda_j\| \, \|R(\omega)\| d\omega$$

$$\geq \int <R(\omega), \Sigma_j a_j \varepsilon_j(\omega)\lambda_j> d\omega$$

$$= \Sigma_j a_j <\int \varepsilon_j(\omega) R(\omega)d\omega, \lambda_j>.$$

But $\int \varepsilon_j(\omega)R_n(\omega)\, d\omega = f_n^j \theta_n = \Delta_n^j$ and thus $\frac{\rho}{4} > \Sigma_j a_j <\Delta^j, \lambda_j>$, a contradiction.

CONSEQUENCES OF THE MAIN RESULT

The following observation will allow us to prove in certain cases the D-P property of a Banach space using the local structure of the space.

PROPOSITION 6 : Let X be a Banach space and assume $X = \overline{\cup_n X_n}$ where (X_n) is an increasing sequence of subspaces of X. If now $\oplus_\infty X_n$ is a D-P space, then X also has D-P property.

Proof : Assume (x_i) a weakly null sequence in X and (x_i^*) a weakly null sequence in X^* such that $<x_i, x_i^*>$ does not tend to null. It is clear that we may assume the x_i in $\cup_n X_n$. Since the X_n are increasing, it is possible to find a subsequence (Y_n) of (X_n) ; so that $x_1, \dots, x_n$ belong to Y_n, for each n. We will show that $\oplus_\infty Y_n$ fails the D-P property. Because $\oplus_\infty Y_n$ is complemented in $\oplus_\infty X_n$, also $\oplus_\infty X_n$ is not D-P, which will complete the proof.

Denote $i_n : Y_n \to X$ the injection and $p_n : \oplus_\infty Y_n \to Y_n$ the projection. Consider the sequence (ξ^i) in $\oplus_\infty Y_n$, where the vector ξ^i is defined by

$$\begin{cases} \xi_n^i = 0 \text{ if } n < i \\ \xi_n^i = x_i \text{ if } n \geq i \end{cases}$$

Let $\mathcal{U}$ be a free ultrafilter on $\mathbb{N}$. Introduce the sequence (η_i) in $[\oplus_\infty Y_n]^*$ by taking $\eta_i = \lim_{\mathcal{U}} p_n^* \, i_n^*(x_i^*)$.

Thus $\langle \xi^i, \eta_i \rangle = \lim_{\mathcal{U}} \langle \xi_n^i, x_i^* \rangle = \langle x_i, x_i^* \rangle$.

It remains to verify that (ξ^i) and (η_i) are weakly null.
So fix an infinite subset M of $\mathbb{N}$ and $\delta > 0$.
Because (x_i) is weakly null, there is a finitely supported sequence $(\lambda_i)_{i \in M}$ of positive scalars such that $\Sigma_i \lambda_i = 1$ and $\|\Sigma_i \lambda_i \varepsilon_i x_i\| \leq \delta$ for all signs $\varepsilon_i = \pm 1$. Clearly $\|\Sigma_i \lambda_i \xi^i\| = \sup_n \|\Sigma_{i \leq n} \lambda_i x_i\| \leq \delta$. Since (x_i^*) is weakly null, there is a convex combination $\Sigma_{i \in M} \lambda_i x_i^*$ so that $\|\Sigma_i \lambda_i x_i^*\| \leq \delta$.
Consequently $\|\Sigma_i \lambda_i p_n^* i_n^*(x_i^*)\| \leq \delta$ for each n and hence also $\|\Sigma_i \lambda_i \eta_i\| < \delta$ as required.

<u>REMARK</u> : Proposition 6 has no converse. Consider for instance the space $X = \oplus_1 l^2(n)$, thus the l^1-sum of the $l^2(n)$-spaces. Then X has the Schur property and hence the D-P property. However, if (X_n) is an increasing sequence of subspaces of X so that $X = \overline{\bigcup_n X_n}$, then $\oplus_\infty X_n$ is never D-P. Indeed, since X_n contain uniformely complemented Hilbert spaces of arbitrarily large dimension, $\oplus_\infty l^2(n)$ is a complemented subspace of $\oplus_\infty X_n$ and $\oplus_\infty l^2(n)$ fails the D-P property by prop. 6 (in fact, l^2 is a complemented subspace of $\oplus_\infty l^2(n)$).

The next result about the local structure of ultraproducts is almost obvious.

<u>LEMMA 7</u> : Let X be a Banach space, m a positive integer, E a finite dimensional Banach space and $\lambda < \infty$. Assume that for any subspace U of X, dim U = m, there exists a subspace V of X such that $U \subset V$ and $d(V,E) \leq \lambda$. Then the same holds for any ultra-product $X_{\mathcal{U}}$ of X.

For any positive integer p, denote E_p the l^∞-sum of p copies of the space $l^1(p)$.

THEOREM 8 : Let X be a Banach space and $\lambda < \infty$ with the following property
For any finite dimensional subspace U of X, there exists a subspace V of X such that $U \subset V$ and $d(V,E_p) \leq \lambda$, for some p.
Then
1. Any ultra-product $X_{\mathcal{U}}$ of X is D-P
2. All duals of X are D-P.

Proof : Using proposition 1.20, the second assertion is clearly a consequence of the first. It is a routine exercice (which is left to the reader) to verify that in fact the following condition is satisfied.
For any integer m, there exists an integer $p = p(m)$ such that if U is a subspace of X, dim U = m, then there is a subspace V of X satisfying $U \subset V$ and $d(V,E_p) \leq \lambda'$ $(\lambda' > \lambda)$.
Now, by lemma 7, any ultra-product $X_{\mathcal{U}}$ has the same property.
Therefore, any separable Y of $X_{\mathcal{U}}$ is contained in a subspace Z of $X_{\mathcal{U}}$ of the form $Z = \overline{\bigcup_n V_n}$, where (V_n) is an increasing sequence of spaces for which $d(V_n,E_{p_n}) \leq \lambda'$, for some sequence (p_n) of integers.
In order to show that $X_{\mathcal{U}}$ is D-P, it is sufficient to prove that each such space Z has D-P property. By prop. 6, it is enough to show that $\oplus_\infty V_n$ is D-P. But $\oplus_\infty V_n$ is isomorphic to $\oplus_\infty E_{p_n}$, which is a complemented subspace of $\oplus_\infty l^1$ and hence of $\oplus_\infty L^1$. So theorem 4 completes the proof.

It is easily seen that C_{L^1} and $(L^1_C)^*$ both satisfy the condition stated in theorem 8 (for details about the dual of L^1_C, we refer the reader to [45]).
So we obtain the following result :

COROLLARY 9 : C_{L^1}, L^1_C and their duals are D-P spaces.

4. REMARKS AND PROBLEMS

1. In fact, theorem 4 is equivalent to the a priori weaker statement that $\oplus_\infty l^1(n)$ is D-P. However, the proof of this result does not seem easier and we also use the "Riesz-product technique".

2. Since, by theorem 4, the space $\oplus_\infty E_p$ has the D-P property, it follows that the E_p do not contain uniformely complemented Hilbert spaces of arbitrarily large dimension. This solves a problem raised in [57], p. 68.

3. It is unknown if in general the D-P property of X implies the D-P property of C_X and L^1_X. Corollary 9 gives us a positive solution to this question in case X is a C or L^1-space.
In fact, we may introduce the sequence $(\mathfrak{X}_n)$ of Banach spaces, taking $\mathfrak{X}_1 = C$

$$\mathfrak{X}_{n+1} = C_{\mathfrak{X}_n} \text{ if n is even}$$

$$\mathfrak{X}_{n+1} = L^1_{\mathfrak{X}_n} \text{ if n is odd.}$$

Using similar techniques, it can be shown that all these spaces (and their duals) are D-P.

4. So far, we do not know the answer to the following question : Suppose (X_n) a sequence of finite dimensional Banach spaces such that $\oplus_\infty X_n$ is D-P. Has $\oplus_\infty X_n^*$ then also the D-P property ?

REFERENCES

[1] D. ALSPACH, P. ENFLO and E. ODELL : On the structure of separable $\mathcal{L}^p$ spaces $(1 < p < \infty)$, Studia Math. 60 (1977), 79-90

[2] Z. ALTSHULER : Characterization of c_0 and ℓ^p among Banach spaces with a symmetric basis, Israel Journal of Math. 24, 39-44 (1976)

[3] Z. ALTSCHULER : A Banach space with a symmetric basis which contains no ℓ^p or c_0, and all its symmetric basic sequences are equivalent, Compositio Math. (1977)

[4] W.G. BADE : On Boolean algebras of projections and algebras of operators, Trans. AMS, 80 (1955), 345-359

[5] W.G. BADE : The Banach space C(S), to appear

[6] S. BANACH : Théorie des opérations linéaires, Warszawa, 1932

[7] S. BANACH and S. MAZUR : Zur Thaorie der linearen Dimension, Studia Math., 4 (1933), 100-112

[8] G. BENNETT, L. DOR, V. GOODMAN, W.B. JOHNSON, C.M. NEWMAN : On uncomplemented subspaces of L^p $(1 < p < 2)$, Israel J. Math. (1977) 178-187

[9] Y. BENYAMINI : Nearly isometric preduals of l^1 which are not isometric, unpublished

[10] Y. BENYAMINI and J. LINDENSTRAUSS : A predual of l^1 which is not isomorphic to a C(K)-space, Israel Journal of Math., 13 (1972), 246-259

[11] C. BESSAGA and A. PELCZYNSKI : On subspaces of a space with an absolute basis, Bull. Acad. Sci. Pol., 6 (1958), 313-314

[12] C. BESSAGA and A. PELCZYNSKI : On bases and unconditional convergence of series in Banach spaces, Studia Math., 17 (1958), 151-164

[13] C. BESSAGA and A. PELCZYNSKI : Spaces of continuous functions IV, Studia Math. 19 (1960) 53-62

[14] J. BOURGAIN : The Szlenk index and operators on C(K)-spaces, Bull. Soc. Math. Belg., Vol 31, Fasc. I, Ser. B, 1979, 87-117

[15] J. BOURGAIN : On separable Banach spaces, universal for all separable reflexive spaces, Proc. AMS, Vol 79, N2, 1980, 241-246

[16] J. BOURGAIN : On the Dunford-Pettis property, Proc. AMS, Vol 81, N2, 1981, 265-272

[17] J. BOURGAIN : A characterization of non-Dunford-Pettis operators on L^1, Israel J. Math., Vol 37, Nos 1-2, 1980, 48-53

[18] J. BOURGAIN : An averaging result for l^1-sequences and applications to weakly conditionally compact sets in L^1_X, Israel Journal Math., Vol 32, No 4, 1979, 289-298

[19] J. BOURGAIN : Remarks on the double dual of a Banach space, Bull. Soc. Math. Belg., to appear

[20] J. BOURGAIN : Dentability and finite dimensional decompositions, Studia Math., T 67 (1980), 135-148

[21] J. BOURGAIN : Un espace $\mathcal{L}^\infty$ jouissant de la propriété de Schur et de la propriété de Radon-Nikodým, Sém. d'Anal. Fonct. 1978-1979, Exp. No IV, Ecole Polytechnique, Centre de Math.

[22] J. BOURGAIN : Un espace non Radon-Nikodým sans arbre diadique, Sém. d'Anal. Fonct. 1978-1979, Exp. No XXIX, Ecole Polytechnique, Centre de Math.

[23] J. BOURGAIN : Dunford-Pettis operators on L^1 and the Radon-Nikodým property, Israel Journal Math., Vol 37, Nos 1-2, 1980, 34-47

[24] J. BOURGAIN : Espaces $\mathcal{L}^1$ ne vérifiant pas la propriété de Radon-Nikodým, C.R. Acad. Sc. Paris, t. 291, Ser. A, 343-345

[25] J. BOURGAIN : A counterexample to a complementation problem, Composition Math., Vol 43, Fasc. 1, 1981, 133-144

[26] J. BOURGAIN : Nouvelles propriétés des espaces L^1/H^1_0 et H^∞, Sém. d'Anal Fonct, 1980-81, Exp. III, Ecole Polytechnique, Centre de Math.

[27] J. BOURGAIN : Some new properties of the Banach spaces L^1/H^1_0 and H^∞, preprint

[28] J. BOURGAIN : A distorted norm on L^1, preprint

[29] J. BOURGAIN and F. DELBAEN : A class of special $\mathcal{L}^\infty$-spaces, Acta Math., Vol 145, 1980, 155-176

[30] J. BOURGAIN, H.P. ROSENTHAL and G. SCHECHTMAN : An ordinal L^p-index for Banach spaces, with application to complemented subspaces of L^p, Annals of Math., to appear

[31] J. BOURGAIN and H.P. ROSENTHAL : Martingales valued in certain subspaces of L^1, Israel J. Math., Vol 37, Nos 1-2, 1980, 54-76

[32] J. BOURGAIN and H.P. ROSENTHAL : Geometrical implications of certain finite dimensional decompositions, Bull. Soc. Math. Belgique, Vol 32, Fasc. 1, Ser B, 1980, 57-82

[33] J. BOURGAIN, D.H. FREMLIN and M. TALAGRAND ; Pointwise compact sets of Baire-measurable functions, Amer. Journal Math. 100 (1978), 845-886

[34] D.L. BURKHOLDER : Martingale transforms, Annals Math. Stat., 37 (1966), 1494-1504

[35] D.L. BURKHOLDER : Distribution function inequalities for martingales, Ann. Prob., 1 (1973), 19-43

[36] D.L. BURKHOLDER, B.J. DAVIS and R.F. GUNDI : Integral inequalities for convex functions of operators on martingales, Proc. of the 6th Berkeley Symp. on Math. Stat. and Prob., (1972), 223-240

[37] D.L. BURKHOLDER and R.F. GUNDI : Extrapolation and interpolation of quasilinear operators on martingales, Acta Math. 124 (1970), 250-299

[38] L. CARLESON : An interpolation problem for bounded analytic functions, Amer. J. Math. 80 (1958), 921-930

[39] P.G. CASAZZA and BOR-LUH·LIN : Projections on Banach spaces with symmetric bases, Studia Math. 52 (1974), 189-193

[40] W.J. DAVIS, T. FIGIEL, W.B. JOHNSON and A. PELCZYNSKI : Factoring weakly compact operators, Journal Funct. Anal. 17 (1974), 311-327

[41] D. DACUNHA-CASTELLE and J.L. KRIVINE : Application des ultraproduits à l'étude des espaces et des algébres de Banach, Studia Math. 41 (1972), 315-334

[42] C. DELLACHERIE : Les dérivations en théorie descriptive des ensembles et le théoréme de la borne, Sém. de Prob. XI, Université de Strasbourg, Lect. Notes, Vol. 581, 34-46, Springer 1977

[43] J. DIESTEL : Geometry of Banach Spaces - Selected Topers, Lect. Notes in Math., Vol 485, Springer 1975

[44] J. DIESTEL and J.J. UHL : The Radon-Nikodým theorem for Banach space valued measures, Rocky Mountain Journal Math., 6 (1976)

[45] J. DIESTEL and J.J. UHL : The theory of vector measures, AMS Surveys, 15 (1977)

[46] L. DOR : On sequences spanning a complex l^1-space, Proc. A.M.S., 47 (1975), 515-516

[47] L. DOR : On projections in L^1, Annals Math. (2) 102, 1975, 463-474

[48] L. DOR : private communication

[49] L. DOR, T. STARBIRD : Projections of L^p onto subspaces spanned by independent random variables, Compositio Math., to appear

[50] N. DUNFORD and J. SCHWARTZ : Linear Operators, Vol I (1958), Interscience

[51] P. DUREN : Theory of H^p Spaces, Academic Press, 1970

[52] A. DVORETZKY and C.A. ROGERS : Absolute and unconditional convergence in normed linear spaces, Proc. Natl. Acad. Sci (USA) 36 (1950), 192-197

[53] I.S. EDELSTEIN : On complemented subspaces and unconditional bases in $l^p \oplus l^2$. Teor. Funkcii. Funct. Anal. i. Prelosen, 10 (1970), 132-143

[54] P. ENFLO : Banach spaces which can be given an equivalent uniformly convex norm, Israel Journal Math., 13 (1972), 281-288

[55] P. ENFLO, H.P. ROSENTHAL : Some results concerning $L^p(\mu)$-spaces, J. Funct. Anal., Vol 14, No 4, 1973, 325-348

[56] P. ENFLO and T. STARBIRD : Subspaces of L^1 containing L^1, to appear in Studia Math. 65

[57] T. FIGIEL, J. LINDENSTRAUSS and V.D. MILMAN : The dimension of almost spherical sections of convex bodies, Acta Math., 139 (1977), 53-94

[58] J. GAMLEN and R. GAUDET : On subsequences of the Haar system in $L_p[0,1]$ $(1 < p < \infty)$, Israel Journal Math., 15 (1973), 404-413

[59] J. GARNETT : Interpolating sequences for bounded harmonic functions, Indiana Univ. Math. J. 21 (1971), 187-192

[60] A.M. GARSIA : Martingale inequalities, Seminar notes on recent progress, Benjamin, Reading, Mass. 1973

[61] Y. GORDON and D.R. LEWIS : Absolutely summing operators and locally unconditional structures, Acta Math., 133 (1974), 27-48

[62] C.C. GRAHAM, O.C. MC GEHEE : Essays in Commutative Harmonic Analysis, Grundlehren der mathematischen Wissenschaffen, 38, Springer Verlag

[63] A. GROTHENDIECK : Sur les applications linéaires faiblement compactes d'espaces du type C(K), Canadian Journal Math. 5 (1953), 129-173

[64] A. GROTHENDIECK : Produits tensoriels topologiques et espaces nucléaires, Memoirs AMS 16 (1955)

[65] S. HEINRICH : Ultraproducts in Banach space Theory, Akademy der Wissenschaften DDR, 1979, preprint

[66] J. HAGLER : A counterexample to several questions about Banach spaces, Studia Math., 60 (1977)

[67] N. HINDMAN : Finite sums from sequences within Cells of a Partition of N, J. of Combinatorial Th. (A) 17, 1974, 1-11

[68] R.C. JAMES : Uniformly non square Banach spaces, Annals Math. 60 (1964), 542-550

[69] R.C. JAMES : Super-reflexive spaces with bases, Pac. Journal Math. 41 (1972), 408-419

[70] W.B. JOHNSON, J. LINDENSTRAUSS : Examples of $\mathcal{L}^1$-spaces, to appear

[71] W.B. JOHNSON, E. ODELL : Subspaces and quotients of $\ell^p \oplus \ell^2$ and X_p, preprint

[72] W.B. JOHNSON and H.P. ROSENTHAL : On w^*-basic sequences and their applications to the study of Banach spaces, Studia Math. 43 (1972), 77-92

[73] W.B. JOHNSON, H.P. ROSENTHAL and M. ZIPPIN : On bases, finitedimensional decompositions and weaker structures in Banach spaces, Israel Journal of Math. 9 (1971), 488-506

[74] W.B. JOHNSON and M. ZIPPIN : On subspaces of quotients of $(\Sigma G_n)_{l^p}$ and $(\Sigma G_n)_{c_0}$, Israel Journal Math., 13 (1972), 311-316

[75] W.B. JOHNSON and M. ZIPPIN : Every separable predual of an L^1-space is a quotient space of $C(\Lambda)$, preprint

[76] W.B. JOHNSON, B. MAUREY, G. SCHECHTMAN and L. TZAFRIRI : Symmetric structures in Banach spaces, Mémoirs of the A.M.S. 217 (1979)

[77] N. KALTON : Embedding of L_1 in a Banach lattice, to appear

[78] M.I. KADEC and A. PELCZYNSKI : Bases, lacunary sequences and complemented subspaces in L^p, Studia Math. 21 (1962) 161-176 NR 27, 2851

[79] S. KAKUTANI : Concrete representation of abstract L-spaces and the mean ergodic theorem, Annals of Math., 42 (1941), 523-537

[80] S. KAKUTANI : Concrete representation of abstract M-spaces, Annals of Math., 42 (1941), 994-1024

[81] Y. KATZNELSON : An introduction to harmonic analysis, Wiley, 1968

[82] J.L. KRIVINE : Sous-espaces et cones convexes dans les espaces L^p, Thèse, Paris 1967

[83] S. KWAPIEN : On operators factorizable through L_p-space, Studia Math., to appear

[84] J. LINDENSTRAUSS and A. PELCZYNSKI : Absolutely summing operators in $\mathcal{L}^p$-spaces and their applications, Studia Math. 29 (1968) 275-326

[85] J. LINDENSTRAUSS and H.P. ROSENTHAL : The $\mathcal{L}^p$-spaces, Israel Journal Math. 7 (1969), 325-349

[86] J. LINDENSTRAUSS and L. TZAFRIRI : Classical Banach spaces I and II, Springer Verlag

[87] J. LINDENSTRAUSS and L. TZAFRIRI : Classical Banach spaces, Lecture Notes, 338 (1973), Springer

[88] J. LINDENSTRAUSS : Extension of compact operators, Mém AMS 48 (1964)

[89] D. LEWIS and C. STEGALL : Banach spaces whose duals are isomorphic to $l^1(\Gamma)$, Journal of Funct. Anal. 12 (1971) 167-177

[90] F. LUST : Ensembles de Rosenthal et ensembles de Riesz, C.R. Acad. Sci. Paris Série A, 282 (1976), 833-835

[91] M. Mc CARTNEY and R. O'BRIEN : preprint

[92] K.R. MILLIKEN : Ramsey's Theorem with Sums or Unions, J. of combinatorial Th. (A) 18, 1975, 276-290

[93] A. MILUTIN : Isomorphisms of spaces of continuous functions on compacta of power continuum, Tieoria Funct. (Kharkov), 2 (1966), 150-156

[94] E. ODELL and H.P. ROSENTHAL : A double dual characterization of separable Banach spaces containing l^1, Israel Journal Math. 20 (1975), 375-384

[95] A. PELCZYNSKI : Projections in certain Banach spaces, Studia Math. 19 (1960), 209-228

[96] A. PELCZYNSKI : Banach spaces in which every unconditionally converging operator is weakly compact, Bull. Acad. Pol. Sci. 12 (1962), 641-648

[97] A. PELCZYNSKI : On C(S) subspaces of separable Banach spaces, Studia Math. 31 (1968), 513-522

[98] A. PELCZYNSKI : Universal bases, Studia Math., 32 (1969), 247-268

[99] A. PELCZYNSKI : Banach spaces of analytic functions and absolutely summing operators, AMS Regional Conference Series Math., 30, Providence, 1977

[100] R.R.PHELPS : Dentability and extreme points in Banach spaces, Journal Funct. Anal. 16 (1974)

[101] G. PISIER : Une propriété de stabilité de la classe des espaces ne contenant pas l^1, C.R. Acad. Sci. Paris, 286 (1978)

[102] G. PISIER : private communication

[103] J.R. RETHERFORD : Operator characterization of $\mathcal{L}^p$-spaces, Israel Journal of Math., 13 (1972), 337-347

[104] H.P. ROSENTHAL : A characterization of Banach spaces containing l^1, Proc. Nat. Acad. Sc. USA 71 (1974), 2411-2413

[105] H.P. ROSENTHAL : Pointwise compact subsets of the first Baire class, Amer. Journal Math., 99, 2 (1977), 362-378

[106] H.P. ROSENTHAL : The Banach spaces C(K) and $L^p(\mu)$, Bull. A.M.S. 81 (1975), 763-781

[107] H.P. ROSENTHAL : Projections onto translation-invariant subspaces of $L^p(G)$, Memoirs AMS, 63

[108] H.P. ROSENTHAL : On injective Banach spaces and the spaces $L^{\infty}(\mu)$ for finite measures μ, Acta Math., 124 (1970), 205-248

[109] H.P. ROSENTHAL : On the subspaces of L^p ($p > 2$) spanned by sequences of independent random variables, Israel Journal Math., 8 (1970), 273-303

[110] H.P. ROSENTHAL : On the span in L^p of sequences of independent random variables (II), Proc. 6th-Berkeley Symp. on Prob. and Stat., Berkeley, Calif. (1971)

[111] H.P. ROSENTHAL : On subspaces of L^p, Annals Math., 17 : 2, 1973, 344-373

[112] H.P. ROSENTHAL : On factors of $C[0,1]$ with non-separable dual, Israel Journal of Math. 13 (1972), 361-378

[113] G. SCHECHTMAN : Examples of $\mathcal{L}^p$-spaces, Israel Journ. of Math. 22 (1975), 138-147

[114] C. STEGALL : The Radon-Nikodým property in conjugate Banach spaces, Trans. A.M.S. 206 (1975), 213-223

[115] J. STERN : Some applications of Model-Theory in Banach space Theory, Ann. Math. Logic 9, 1976, 49-122

[116] E. STEIN : Topics in harmonic analysis related to the Littlewood-Paley theory, Annals of Math. Studies, Princeton 1970

[117] N. Th. VAROPOULOS : Sur un probléme d'interpolation, C.R. Acad. Sc. Paris, Su A, t. 274 (1972), 1905-1908

[118] M. ZIPPIN : On some subspaces of Banach spaces whose duals are L^1-spaces, Proc. AMS 23 (1969), 378-385

SUBJECT INDEX

Vol. 728: Non-Commutative Harmonic Analysis. Proceedings, 1978. Edited by J. Carmona and M. Vergne. V, 244 pages. 1979.

Vol. 729: Ergodic Theory. Proceedings, 1978. Edited by M. Denker and K. Jacobs. XII, 209 pages. 1979.

Vol. 730: Functional Differential Equations and Approximation of Fixed Points. Proceedings, 1978. Edited by H.-O. Peitgen and H.-O. Walther. XV, 503 pages. 1979.

Vol. 731: Y. Nakagami and M. Takesaki, Duality for Crossed Products of von Neumann Algebras. IX, 139 pages. 1979.

Vol. 732: Algebraic Geometry. Proceedings, 1978. Edited by K. Lønsted. IV, 658 pages. 1979.

Vol. 733: F. Bloom, Modern Differential Geometric Techniques in the Theory of Continuous Distributions of Dislocations. XII, 206 pages. 1979.

Vol. 734: Ring Theory, Waterloo, 1978. Proceedings, 1978. Edited by D. Handelman and J. Lawrence. XI, 352 pages. 1979.

Vol. 735: B. Aupetit, Propriétés Spectrales des Algèbres de Banach. XII, 192 pages. 1979.

Vol. 736: E. Behrends, M-Structure and the Banach-Stone Theorem. X, 217 pages. 1979.

Vol. 737: Volterra Equations. Proceedings 1978. Edited by S.-O. Londen and O. J. Staffans. VIII, 314 pages. 1979.

Vol. 738: P. E. Conner, Differentiable Periodic Maps. 2nd edition, IV, 181 pages. 1979.

Vol. 739: Analyse Harmonique sur les Groupes de Lie II. Proceedings, 1976–78. Edited by P. Eymard et al. VI, 646 pages. 1979.

Vol. 740: Séminaire d'Algèbre Paul Dubreil. Proceedings, 1977–78. Edited by M.-P. Malliavin. V, 456 pages. 1979.

Vol. 741: Algebraic Topology, Waterloo 1978. Proceedings. Edited by P. Hoffman and V. Snaith. XI, 655 pages. 1979.

Vol. 742: K. Clancey, Seminormal Operators. VII, 125 pages. 1979.

Vol. 743: Romanian-Finnish Seminar on Complex Analysis. Proceedings, 1976. Edited by C. Andreian Cazacu et al. XVI, 713 pages. 1979.

Vol. 744: I. Reiner and K. W. Roggenkamp, Integral Representations. VIII, 275 pages. 1979.

Vol. 745: D. K. Haley, Equational Compactness in Rings. III, 167 pages. 1979.

Vol. 746: P. Hoffman, τ-Rings and Wreath Product Representations. V, 148 pages. 1979.

Vol. 747: Complex Analysis, Joensuu 1978. Proceedings, 1978. Edited by I. Laine, O. Lehto and T. Sorvali. XV, 450 pages. 1979.

Vol. 748: Combinatorial Mathematics VI. Proceedings, 1978. Edited by A. F. Horadam and W. D. Wallis. IX, 206 pages. 1979.

Vol. 749: V. Girault and P.-A. Raviart, Finite Element Approximation of the Navier-Stokes Equations. VII, 200 pages. 1979.

Vol. 750: J. C. Jantzen, Moduln mit einem höchsten Gewicht. III, 195 Seiten. 1979.

Vol. 751: Number Theory, Carbondale 1979. Proceedings. Edited by M. B. Nathanson. V, 342 pages. 1979.

Vol. 752: M. Barr, *-Autonomous Categories. VI, 140 pages. 1979.

Vol. 753: Applications of Sheaves. Proceedings, 1977. Edited by M. Fourman, C. Mulvey and D. Scott. XIV, 779 pages. 1979.

Vol. 754: O. A. Laudal, Formal Moduli of Algebraic Structures. III, 161 pages. 1979.

Vol. 755: Global Analysis. Proceedings, 1978. Edited by M. Grmela and J. E. Marsden. VII, 377 pages. 1979.

Vol. 756: H. O. Cordes, Elliptic Pseudo-Differential Operators – An Abstract Theory. IX, 331 pages. 1979.

Vol. 757: Smoothing Techniques for Curve Estimation. Proceedings, 1979. Edited by Th. Gasser and M. Rosenblatt. V, 245 pages. 1979.

Vol. 758: C. Năstăsescu and F. Van Oystaeyen; Graded and Filtered Rings and Modules. X, 148 pages. 1979.

Vol. 759: R. L. Epstein, Degrees of Unsolvability: Structure and Theory. XIV, 216 pages. 1979.

Vol. 760: H.-O. Georgii, Canonical Gibbs Measures. VIII, 190 pages. 1979.

Vol. 761: K. Johannson, Homotopy Equivalences of 3-Manifolds with Boundaries. 2, 303 pages. 1979.

Vol. 762: D. H. Sattinger, Group Theoretic Methods in Bifurcation Theory. V, 241 pages. 1979.

Vol. 763: Algebraic Topology, Aarhus 1978. Proceedings, 1978. Edited by J. L. Dupont and H. Madsen. VI, 695 pages. 1979.

Vol. 764: B. Srinivasan, Representations of Finite Chevalley Groups. XI, 177 pages. 1979.

Vol. 765: Padé Approximation and its Applications. Proceedings, 1979. Edited by L. Wuytack. VI, 392 pages. 1979.

Vol. 766: T. tom Dieck, Transformation Groups and Representation Theory. VIII, 309 pages. 1979.

Vol. 767: M. Namba, Families of Meromorphic Functions on Compact Riemann Surfaces. XII, 284 pages. 1979.

Vol. 768: R. S. Doran and J. Wichmann, Approximate Identities and Factorization in Banach Modules. X, 305 pages. 1979.

Vol. 769: J. Flum, M. Ziegler, Topological Model Theory. X, 151 pages. 1980.

Vol. 770: Séminaire Bourbaki vol. 1978/79 Exposés 525–542. IV, 341 pages. 1980.

Vol. 771: Approximation Methods for Navier-Stokes Problems. Proceedings, 1979. Edited by R. Rautmann. XVI, 581 pages. 1980.

Vol. 772: J. P. Levine, Algebraic Structure of Knot Modules. XI, 104 pages. 1980.

Vol. 773: Numerical Analysis. Proceedings, 1979. Edited by G. A. Watson. X, 184 pages. 1980.

Vol. 774: R. Azencott, Y. Guivarc'h, R. F. Gundy, Ecole d'Eté de Probabilités de Saint-Flour VIII-1978. Edited by P. L. Hennequin. XIII, 334 pages. 1980.

Vol. 775: Geometric Methods in Mathematical Physics. Proceedings, 1979. Edited by G. Kaiser and J. E. Marsden. VII, 257 pages. 1980.

Vol. 776: B. Gross, Arithmetic on Elliptic Curves with Complex Multiplication. V, 95 pages. 1980.

Vol. 777: Séminaire sur les Singularités des Surfaces. Proceedings, 1976-1977. Edited by M. Demazure, H. Pinkham and B. Teissier. IX, 339 pages. 1980.

Vol. 778: SK_1 von Schiefkörpern. Proceedings, 1976. Edited by P. Draxl and M. Kneser. II, 124 pages. 1980.

Vol. 779: Euclidean Harmonic Analysis. Proceedings, 1979. Edited by J. J. Benedetto. III, 177 pages. 1980.

Vol. 780: L. Schwartz, Semi-Martingales sur des Variétés, et Martingales Conformes sur des Variétés Analytiques Complexes. XV, 132 pages. 1980.

Vol. 781: Harmonic Analysis Iraklion 1978. Proceedings 1978. Edited by N. Petridis, S. K. Pichorides and N. Varopoulos. V, 213 pages. 1980.

Vol. 782: Bifurcation and Nonlinear Eigenvalue Problems. Proceedings, 1978. Edited by C. Bardos, J. M. Lasry and M. Schatzman. VIII, 296 pages. 1980.

Vol. 783: A. Dinghas, Wertverteilung meromorpher Funktionen in ein- und mehrfach zusammenhängenden Gebieten. Edited by R. Nevanlinna and C. Andreian Cazacu. XIII, 145 pages. 1980.

Vol. 784: Séminaire de Probabilités XIV. Proceedings, 1978/79. Edited by J. Azéma and M. Yor. VIII, 546 pages. 1980.

Vol. 785: W. M. Schmidt, Diophantine Approximation. X, 299 pages. 1980.

Vol. 786: I. J. Maddox, Infinite Matrices of Operators. V, 122 pages. 1980.

Vol. 787: Potential Theory, Copenhagen 1979. Proceedings, 1979. Edited by C. Berg, G. Forst and B. Fuglede. VIII, 319 pages. 1980.

Vol. 788: Topology Symposium, Siegen 1979. Proceedings, 1979. Edited by U. Koschorke and W. D. Neumann. VIII, 495 pages. 1980.

Vol. 789: J. E. Humphreys, Arithmetic Groups. VII, 158 pages. 1980.

Vol. 790: W. Dicks, Groups, Trees and Projective Modules. IX, 127 pages. 1980.

Vol. 791: K. W. Bauer and S. Ruscheweyh, Differential Operators for Partial Differential Equations and Function Theoretic Applications. V, 258 pages. 1980.

Vol. 792: Geometry and Differential Geometry. Proceedings, 1979. Edited by R. Artzy and I. Vaisman. VI, 443 pages. 1980.

Vol. 793: J. Renault, A Groupoid Approach to C*-Algebras. III, 160 pages. 1980.

Vol. 794: Measure Theory, Oberwolfach 1979. Proceedings 1979. Edited by D. Kölzow. XV, 573 pages. 1980.

Vol. 795: Séminaire d'Algèbre Paul Dubreil et Marie-Paule Malliavin. Proceedings 1979. Edited by M. P. Malliavin. V, 433 pages. 1980.

Vol. 796: C. Constantinescu, Duality in Measure Theory. IV, 197 pages. 1980.

Vol. 797: S. Mäki, The Determination of Units in Real Cyclic Sextic Fields. III, 198 pages. 1980.

Vol. 798: Analytic Functions, Kozubnik 1979. Proceedings. Edited by J. Ławrynowicz. X, 476 pages. 1980.

Vol. 799: Functional Differential Equations and Bifurcation. Proceedings 1979. Edited by A. F. Izé. XXII, 409 pages. 1980.

Vol. 800: M.-F. Vignéras, Arithmétique des Algèbres de Quaternions. VII, 169 pages. 1980.

Vol. 801: K. Floret, Weakly Compact Sets. VII, 123 pages. 1980.

Vol. 802: J. Bair, R. Fourneau, Etude Géometrique des Espaces Vectoriels II. VII, 283 pages. 1980.

Vol. 803: F.-Y. Maeda, Dirichlet Integrals on Harmonic Spaces. X, 180 pages. 1980.

Vol. 804: M. Matsuda, First Order Algebraic Differential Equations. VII, 111 pages. 1980.

Vol. 805: O. Kowalski, Generalized Symmetric Spaces. XII, 187 pages. 1980.

Vol. 806: Burnside Groups. Proceedings, 1977. Edited by J. L. Mennicke. V, 274 pages. 1980.

Vol. 807: Fonctions de Plusieurs Variables Complexes IV. Proceedings, 1979. Edited by F. Norguet. IX, 198 pages. 1980.

Vol. 808: G. Maury et J. Raynaud, Ordres Maximaux au Sens de K. Asano. VIII, 192 pages. 1980.

Vol. 809: I. Gumowski and Ch. Mira, Recurences and Discrete Dynamic Systems. VI, 272 pages. 1980.

Vol. 810: Geometrical Approaches to Differential Equations. Proceedings 1979. Edited by R. Martini. VII, 339 pages. 1980.

Vol. 811: D. Normann, Recursion on the Countable Functionals. VIII, 191 pages. 1980.

Vol. 812: Y. Namikawa, Toroidal Compactification of Siegel Spaces. VIII, 162 pages. 1980.

Vol. 813: A. Campillo, Algebroid Curves in Positive Characteristic. V, 168 pages. 1980.

Vol. 814: Séminaire de Théorie du Potentiel, Paris, No. 5. Proceedings. Edited by F. Hirsch et G. Mokobodzki. IV, 239 pages. 1980.

Vol. 815: P. J. Slodowy, Simple Singularities and Simple Algebraic Groups. XI, 175 pages. 1980.

Vol. 816: L. Stoica, Local Operators and Markov Processes. VIII, 104 pages. 1980.

Vol. 817: L. Gerritzen, M. van der Put, Schottky Groups and Mumforc Curves. VIII, 317 pages. 1980.

Vol. 818: S. Montgomery, Fixed Rings of Finite Automorphism Groups of Associative Rings. VII, 126 pages. 1980.

Vol. 819: Global Theory of Dynamical Systems. Proceedings, 1979. Edited by Z. Nitecki and C. Robinson. IX, 499 pages. 1980.

Vol. 820: W. Abikoff, The Real Analytic Theory of Teichmüller Space VII, 144 pages. 1980.

Vol. 821: Statistique non Paramétrique Asymptotique. Proceedings, 1979. Edited by J.-P. Raoult. VII, 175 pages. 1980.

Vol. 822: Séminaire Pierre Lelong–Henri Skoda, (Analyse) Années 1978/79. Proceedings. Edited by P. Lelong et H. Skoda. VIII, 356 pages, 1980.

Vol. 823: J. Král, Integral Operators in Potential Theory. III, 171 pages. 1980.

Vol. 824: D. Frank Hsu. Cyclic Neofields and Combinatorial Designs. VI, 230 pages. 1980.

Vol. 825: Ring Theory, Antwerp 1980. Proceedings. Edited by F. van Oystaeyen. VII, 209 pages. 1980.

Vol. 826: Ph. G. Ciarlet et P. Rabier, Les Equations de von Kármán. VI, 181 pages. 1980.

Vol. 827: Ordinary and Partial Differential Equations. Proceedings, 1978. Edited by W. N. Everitt. XVI, 271 pages. 1980.

Vol. 828: Probability Theory on Vector Spaces II. Proceedings, 1979. Edited by A. Weron. XIII, 324 pages. 1980.

Vol. 829: Combinatorial Mathematics VII. Proceedings, 1979. Edited by R. W. Robinson et al.. X, 256 pages. 1980.

Vol. 830: J. A. Green, Polynomial Representations of GL_n. VI, 118 pages. 1980.

Vol. 831: Representation Theory I. Proceedings, 1979. Edited by V. Dlab and P. Gabriel. XIV, 373 pages. 1980.

Vol. 832: Representation Theory II. Proceedings, 1979. Edited by V. Dlab and P. Gabriel. XIV, 673 pages. 1980.

Vol. 833: Th. Jeulin, Semi-Martingales et Grossissement d'une Filtration. IX, 142 Seiten. 1980.

Vol. 834: Model Theory of Algebra and Arithmetic. Proceedings, 1979. Edited by L. Pacholski, J. Wierzejewski, and A. J. Wilkie. VI, 410 pages. 1980.

Vol. 835: H. Zieschang, E. Vogt and H.-D. Coldewey, Surfaces and Planar Discontinuous Groups. X, 334 pages. 1980.

Vol. 836: Differential Geometrical Methods in Mathematical Physics. Proceedings, 1979. Edited by P. L. García, A. Pérez-Rendón, and J. M. Souriau. XII, 538 pages. 1980.

Vol. 837: J. Meixner, F. W. Schäfke and G. Wolf, Mathieu Functions and Spheroidal Functions and their Mathematical Foundations Further Studies. VII, 126 pages. 1980.

Vol. 838: Global Differential Geometry and Global Analysis. Proceedings 1979. Edited by D. Ferus et al. XI, 299 pages. 1981.

Vol. 839: Cabal Seminar 77 – 79. Proceedings. Edited by A. S. Kechris, D. A. Martin and Y. N. Moschovakis. V, 274 pages. 1981.

Vol. 840: D. Henry, Geometric Theory of Semilinear Parabolic Equations. IV, 348 pages. 1981.

Vol. 841: A. Haraux, Nonlinear Evolution Equations- Global Behaviour of Solutions. XII, 313 pages. 1981.

Vol. 842: Séminaire Bourbaki vol. 1979/80. Exposés 543–560. IV, 317 pages. 1981.

Vol. 843: Functional Analysis, Holomorphy, and Approximation Theory. Proceedings. Edited by S. Machado. VI, 636 pages. 1981.